HOW TO PASS

INTERMEDIATE 2
CHEMISTRY

Martin Armitage

Hodder Gibson

A MEMBER OF THE HODDER HEADLINE GROUP

Acknowledgements

Photo credits
Figure 2.1 CartoonStock; Figure 2.5 © Roman Krochuk / iStockphoto.com; Figure 3.2 © Floortje/istockphoto.com; Figure 3.4 © Edyta Pawłowska / iStockphoto.com; Figure 3.5 © Holmes Garden Photos / Alamy; Figure 3.6 © Christopher Dodson / iStockphoto.com; Figure 4.1 NASA/SCIENCE PHOTO LIBRARY; Figure 4.2 CartoonStock; Figure 4.3 John Cancalosi/Still Pictures; Figure 4.4 JA Electronics Mfg. Co., Inc.

Every effort has been made to trace all copyright holders, but if any have been inadvertently overlooked the Publishers will be pleased to make the necessary arrangements at the first opportunity.

Although every effort has been made to ensure that website addresses are correct at time of going to press, Hodder Gibson cannot be held responsible for the content of any website mentioned in this book. It is sometimes possible to find a relocated web page by typing in the address of the home page for a website in the URL window of your browser.

Hachette's policy is to use papers that are natural, renewable and recyclable products and made from wood grown in sustainable forests. The logging and manufacturing processes are expected to conform to the environmental regulations of the country of origin.

Orders: please contact Bookpoint Ltd, 130 Milton Park, Abingdon, Oxon OX14 4SB. Telephone: (44) 01235 827720. Fax: (44) 01235 400454. Lines are open 9.00 – 5.00, Monday to Saturday, with a 24-hour message answering service. Visit our website at www.hoddereducation.co.uk. Hodder Gibson can be contacted direct on:
Tel: 0141 848 1609; Fax: 0141 889 6315; email: hoddergibson@hodder.co.uk

© Martin Armitage 2008
First published in 2008 by
Hodder Gibson, an imprint of Hodder Education, part of Hachette Livre UK,
2a Christie Street
Paisley PA1 1NB

Impression number   5 4 3 2 1
Year   2012 2011 2010 2009 2008

Cover photo MARK SYKES/SCIENCE PHOTO LIBRARY
Illustrations by Richard Duszczak, Cartoon Studio Limited
Typeset in 9.5 on 12.5pt Frutiger Light by Phoenix Photosetting, Chatham, Kent
Printed in Great Britain by Martins The Printers, Berwick-upon-Tweed

A catalogue record for this title is available from the British Library

ISBN-13: 978 0340 946 336

CONTENTS

DEDICATION

This book is dedicated to the memory of Norman Conquest, a valued friend and fellow chemistry teacher.

INTRODUCTION

Congratulations – you've decided to try Intermediate 2 Chemistry!

You've chosen a subject which is very useful in its own right but is also an excellent springboard for Higher Chemistry, if you decide to go on with it. It will give you a great understanding of the world around you, including environmental and industrial issues. You'll also learn skills that are useful in areas other than chemistry – 'transferable skills' like numeracy and problem solving. Maybe you've done Standard Grade Chemistry and now you're doing Intermediate 2 aiming to do Higher in sixth year – you should find Intermediate 2 quite easy as you'll have met a lot of the chemistry before.

You'll find suggestions about how to revise in this book. You'll also find short summaries of each topic and examples of questions with discussions of how to tackle them. However, this isn't a textbook – it's a book about what to do to pass Intermediate 2 Chemistry. The most important factors in passing an exam like this are your approach to revision and the amount of work you do. This book isn't here to replace your teachers, your class notes or your textbook. It's here to help you to make the very best use of these.

Good luck!

Martin Armitage

HOW TO REVISE

In the Intermediate 2 exam, more than 60% of the marks can be for Knowledge and Understanding. If you've already done Standard Grade Chemistry you'll remember that only 40% of the overall marks were for Knowledge and Understanding.

The message is simple – you really have to learn your stuff to do well in Intermediate 2.

What's in the course?

So, what will you have to revise?

The course consists of three units, like most National Qualifications courses.

◆ **Unit 1 – Building Blocks**

◆ **Unit 2 – Carbon Compounds**

◆ **Unit 3 – Acids, Bases and Metals**

You can look forward to a NAB at the end of each unit. You'll have to score 18 out of 30 to pass the NAB (but don't worry – half marks are rounded upward!) and you'll have to pass all three NABs to get an award in Intermediate 2 Chemistry at the end of the course – assuming, of course, that you pass the final exam. NAB questions are pretty easy compared with questions in the final exam, so don't get the idea that you've as good as passed the final exam when you sail through the first NAB. Wait till you do your Prelim!

Each unit is made up of several topics – altogether there are fourteen topics in the course.

The final exam tests every topic, so you'll have to revise the whole course – there's no point in trying to guess if something won't be tested because it all will.

Getting organised

Your course is probably taught topic by topic, with handouts, worksheets or notes to back each topic up. You'll probably be using a textbook as well. So by the end of your course you'll have a heap of paper. Don't let it get out of control. It's best to be organised right from the start, rather than trying to sort through a great heap of papers later on. Different methods suit different people but a cheap and easy way

Figure 1.1 Make sure your notes are well organised

is to buy a box of transparent polythene pockets and to keep each handout or sheet in one of its own. Use a separate ring binder for each unit – that's three ring binders and some poly pockets.

Whenever you receive a sheet or a handout, date it right away. This will let you keep all the sheets in the right order. It's a great idea to read over your notes the evening of the day you get them – to make sure that you actually can read them (it helps!) and that you know what they're about. If in doubt, get help at your next visit to chemistry. It's best to do this when the lesson is fresh in your mind, and it makes learning a lot easier.

You might receive summary notes from time to time. But why not make your own at the end of each topic? It helps you to understand what you've been learning, and you'll learn faster from summaries you've made yourself. And if you don't understand it you won't be able to write a summary, so that's a signal to get help from your teacher.

When should I start?

Start revising when you start the course. OK, you have other homework and it can be difficult to fit in additional revision. The very least you should do is to revise the **previous** lesson in chemistry before the **next** lesson. That makes the new work much easier, and you'll find out what you didn't understand from the last lesson so you can get help.

What should I revise?

No problem here! You revise *everything*, because *everything* is likely to be examined. If you've already got a pass in Standard Grade then you'll find that you've met quite a lot of the chemistry before, and that should make revision easier. But don't feel that you should revise your Standard Grade work as well – it just isn't necessary.

How should I revise?

Most people find that they learn more efficiently when they are interacting with the work. This means that you shouldn't just sit with your notes in front of you, reading them and hoping that something will stick. Have paper and pencil in front of you.

Suppose that you're trying to learn an organic structure – look at it in your notes, then cover it and see if you can write it down correctly on your paper. If you've read over a paragraph of notes, see if you can write down the key points of what you've read.

You can make **flash cards**. Suppose you want to learn the different functional groups, like hydroxyl (– OH) and carboxyl (– COOH). You can write the name of the group on one side of a small piece of paper or card and the actual structure on the other. Like this:

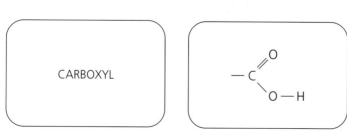

This means that you can test your knowledge of the structures from the names, and also the names from the structures. You can get other people to check your knowledge – even if they don't know any chemistry. You can use flash cards for all sorts of things – you can use them to revise chemical tests, like this:

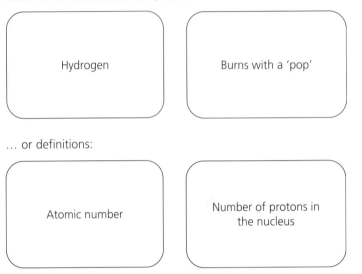

Hydrogen

Burns with a 'pop'

… or definitions:

Atomic number

Number of protons in the nucleus

Mnemonics

A mnemonic is a handy way of learning the names of the members of a set of substances. For example, 'I bring clay for our new home' helps you to you to remember the names of the diatomic elements – iodine, bromine, chlorine, fluorine, oxygen, nitrogen and hydrogen.

And 'Mice eat peanut butter plus ham here on nice days' will help you to remember something else – you'll find a clue on page 6 of the Data Booklet!

Mental maps

Mental maps are a great way of organising information. Here's a chunk taken from the Intermediate 2 Arrangements information – you'll find this on the Internet at www.sqa.org.uk

Sub-atomic particles
Every element is made up of very small particles called atoms.

The atom has a nucleus, which contains protons and neutrons, with electrons moving around outside the nucleus.

Protons have a charge of one-positive, neutrons are neutral and electrons have a charge of one-negative.

Important numbers
Atoms of different elements have a different number of protons, called the atomic number.

The electrons in an atom are arranged in energy levels.

The elements of the Periodic Table are arranged in terms of their atomic number and chemical properties.

Elements with the same number of outer electrons have similar chemical properties.

An atom has a mass number which equals the number of protons added to the number of neutrons.

Now look at the 'mental map' shown in Figure 1.2. See how it links the information. You'd do this once you had covered the topics in class, of course, not from the Intermediate 2 Arrangements.

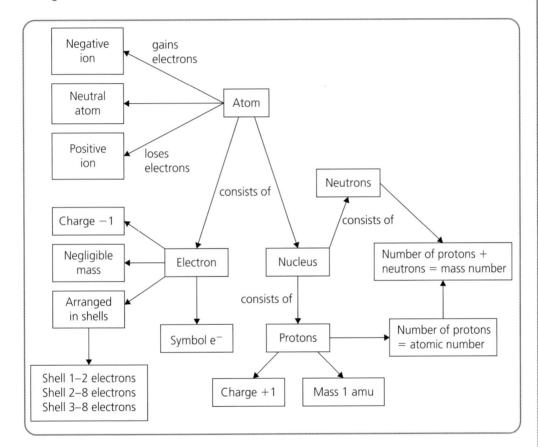

Figure 1.2 A mental map for atomic structure

There's no right or wrong way of making a mental map. It's how *you* see the information that matters.

Glossaries

Try making a glossary – that's a list of terms with their meanings. Then get someone else to help by asking you questions from your glossary. The great thing is that it needn't be someone who knows anything about chemistry!

The Internet

There are some very useful websites which will help you with your revision, especially:

♦ www.evans2chemweb.co.uk

♦ www.bbc.co.uk/revision – 'Bitesize' haven't yet set up Intermediate 2 chemistry revision, though you might find bits of the Standard Grade and Higher Bitesize useful.

♦ www.sqa.org.uk – the SQA website. Here, you'll find the official marking instructions for the most recent Intermediate 2 exams. You'll also find the Arrangements for the exam. These tell you what's in your course so you can check the progress of your revision. Print out the section of the Arrangements on course content and use it as a checklist for revision. It also gives you an idea of the pace of your revision and it's handy for making a note of things you don't understand, so you can ask your teacher or use one of the websites to get help.

Routine

Regular revision is best, although it's hard to do this until the final exams are coming up when you are likely to be on study leave. At other times you have other homework to do. So you might try to fit in a couple of half-hour sessions after you've done your set homework each evening (not Friday of course!) where you revise subjects in which you don't have set homework. Later on, when your courses are complete and you're doing mainly revision at school, you can make a timetable along the following lines.

Monday		Tuesday		Wednesday		Thursday	
6p.m.	Chemistry	6p.m.	English	6p.m.	French	6p.m.	Biology
7p.m.	English	7p.m.	French	7p.m.	Biology	7p.m.	Chemistry
8.15p.m.	French	8.15p.m.	Biology	8.15p.m.	Chemistry	8.15p.m.	English
9.15p.m.	Biology	9.15p.m.	Chemistry	9.15p.m.	English	9.15p.m.	French

Each slot is 50 minutes, so you can have longer or shorter breaks. If you don't like to always follow biology with chemistry, or some other combination, you can shuffle them more. The point is that it can suit some people to always follow the same routine – others like to vary it a bit.

The example above shows only four evenings. It's important to have time off where you can do something quite different and be refreshed for the next set of revision days.

What's the exam like?

Your exam will be for 80 marks. 30 are for multiple-choice questions. In these you select one correct answer from four choices.

If you pick more than one answer, then you'll get no marks for that question.

The answers will be divided roughly equally among As, Bs, Cs and Ds, so don't expect all the answers to be A (or B or C or D for that matter!). Don't leave out a question if you don't know the answer. Any guess is better than no guess! You have one chance in four of being correct if you guess (better odds than the National Lottery and a lot better than your chances of being struck by lightning).

Fifty of the marks are for 'extended response questions' – where you have to write the answer. Most of them are for 1 mark only and need only short answers – a word (or two) or a structure or formula.

Look at the wording of the question. If it says 'Name the …' or 'Draw the structure of …' that's all you have to do – name a compound or process or draw a structure.

Figure 1.3 The chance of a correct guess in a multi-choice question is better than of this happening to you!

There are some questions for 2 marks. They're usually calculations or involve drawing graphs (more about these questions later) or questions which involve completing a table of data.

The Data Booklet

Know your way around the booklet – it can help you in lots of ways. Don't forget that page 4 shows the formulae and charges of ions containing more than one atom, so you shouldn't find it hard to write the formula for any compound, if required.

On page 6, there's a table which tells you about the alkanes and alkenes. They're given *in order of increasing number of carbon atoms*. The third one down is propane. That tells you that propane has three carbon atoms, and so on. You'll never have an excuse for getting the number of carbon atoms wrong in an organic compound.

On the same page there's a table which gives information about alcohols and acids. This is helpful if your mind goes blank when you're trying to answer a question in which you have to name such a compound, because you'll find the names of all the common alcohols and acids there. Again, they're in order of increasing number of carbon atoms, so you should be able to spot that butanoic acid (for example) has four carbon atoms. And don't forget that when you're naming an alcohol, you may have to put in a number to show where the – OH group is. If you call propan-2-ol simply propanol then you might lose half a mark.

Chapter 2

BUILDING BLOCKS

This unit introduces atoms and different kinds of substances and shows why these behave in different ways. It takes a look at how chemical changes take place and also at factors that affect how quickly the changes take place. It also introduces the chemist's way of measuring amounts of substances – the mole.

The unit covers the following topics.

◆ Substances

◆ Reaction rates

◆ The structure of the atom

◆ Bonding, structure and properties

◆ Chemical symbolism

◆ The mole

Substances

Summary

There are about 100 elements, each with its own symbol. A symbol is a shorthand way of identifying the element. They are all shown in the *Periodic Table*. Some of the symbols are really just abbreviations of the name, like Mg for magnesium. Others are taken from old names for the elements, like Fe for iron.

If you want to indicate whether an element is a solid or a liquid or a gas you use a *state symbol* after the symbol. Mg(s) means solid magnesium, Hg(l) means liquid mercury and (g) is used for gases.

A horizontal row of the table is called a *period* – the vertical columns are called *groups*. Elements in the same group have similar chemical properties. Important groups include:

Figure 2.1 Chemists need to be able to recognise symbols

Summary continued ➣

Summary *continued*

◆ Group 1 – the *alkali metals*, a family of very reactive metals;

◆ Group 7 – the *halogens*, a family of very reactive non-metals;

◆ Group 0 – the *noble gases*, a family of very unreactive gases.

The *transition metals* are found between Groups 2 and 3.

Elements combine together to form *compounds*. The name of the compound often indicates what elements are in it. For example, copper sulphide has just copper and sulphur – '*-ide*' tells us there are just two elements. Copper sulphate has three elements, obviously copper and sulphur – and '*-ate*' tells us that there is oxygen in it as well. Copper sulphite also contains copper, sulphur and oxygen – but less oxygen than there is in copper sulphate.

When substances react, new substances form and there is usually an energy change – energy may be given out (*exothermic*) or taken in (*endothermic*).

When you put substances together and they don't react, you are left with a *mixture*. Air is a mixture of about 78% nitrogen, 21% oxygen, 0.9% argon and 0.03% carbon dioxide. You probably know that the test for oxygen is that it relights a glowing splint, and that the percentage of oxygen in the air isn't large enough to let this happen in air.

Solutions are liquid mixtures. The *solvent* is the liquid that does the dissolving – the substance that dissolves is called the *solute*. If there's just a little solute dissolved in the solution then we have a *dilute* solution. If there's a lot of solute dissolved in the solution we have a *concentrated* solution. If so much solute has dissolved in the solution that no more will dissolve then we have a *saturated* solution. We use the state symbol (aq) to show that a substance is dissolved in water.

Example 1

Which of the following compounds contains both a transition metal ion and a halide ion?

A   Aluminium bromide

B   Cobalt chloride

C   Iron oxide

D   Sodium fluoride

Solution

This really shouldn't cause too much of a problem. You just need to be able to spot the names of transition metals and halogens (or a name that looks as if it's got something to do with halogens!). If in doubt about transition metals, check out the Data Booklet (page 8). You'll find that cobalt and iron are both transition metals. So the answer must be either B or C. How about the halide ion? 'Chloride' looks as if it has something to do with chlorine, which is a halogen. Note that you're expected to know that the halogens are Group 7 – the Data Booklet doesn't tell you this! So **B** is the answer.

Example 2

Which of the following gases is a noble gas?

A   Nitrogen

B   Fluorine

C   Oxygen

D   Neon

Solution

There's not much to answering this question, but you're expected to know that the noble gases are the Group 0 elements. The Data Booklet doesn't tell you this. Neon is in Group 0 so the answer is **D**.

Example 3

Which of the following compounds contains only two elements?

A   Calcium hydroxide

B   Calcium phosphate

C   Calcium sulphite

D   Calcium nitride

Solution

You're expected to know that the '-ide' ending tells you that there are only two elements in the compound. But watch out because 'hydroxide' contains both oxygen and hydrogen, so A actually has three elements! The '-ate' and '-ite' endings of B and C indicate that these compounds also contain oxygen, so they contain three elements not two. The answer is **D**, which contains only calcium and nitrogen.

Reaction rates

Summary

In a chemical reaction, the *reactants* (the chemicals you start with) gradually turn into *products*. So, the quantity of reactants present gets less and less as time passes, while the quantity of products gets more and more. You can work out the *rate* of a reaction by measuring the *concentration* of one of the reactants at a certain stage of the reaction and again after some time has passed. You calculate the rate of the reaction by dividing the change in concentration by the time passed:

Summary continued ➤

Summary *continued*

$$\text{rate} = \frac{\text{change in concentration}}{\text{time taken}}$$

You can use the mass of reactant or volume of reactant (especially if it's a gas) as a measure of concentration. You can also use the concentration of one of the products to calculate the rate.

Sometimes, you can calculate the rate simply by working out the reciprocal of the time taken for the reaction:

$$\text{rate} = \frac{1}{\text{time taken}}$$

The rate is always 'something' divided by time, so its units always involve time^{-1}.

You can speed up a reaction by increasing the concentration of the reactants. There are then more particles (atoms or molecules) and they collide with each other more often, resulting in more reactions between them. In a similar way, the reaction is faster if you use a fine powder instead of big lumps of chemical – there is a bigger overall surface area for particles to collide with. Another way of increasing the rate of a reaction is to heat it up.

Catalysts

You can speed up a reaction by adding a *catalyst*. The great thing about a catalyst is that it isn't used up in the reaction. If the catalyst is in the same physical state as the reactants then it's called a *homogeneous* catalyst. If it's in a different state then it's *heterogeneous*.

Heterogeneous catalysts are usually solids and they work by having 'active sites' on their surface. The reactant molecules are adsorbed on to the surface where they react more easily than they would if there were no catalyst present. Catalyst *poisons* act by being adsorbed more tightly than the reactant molecules, so the active sites are all occupied and there's no space for reactant molecules. In industry it's often necessary to replace or regenerate solid catalysts because they gradually become poisoned. The catalysts in the catalytic converter of a car are heterogeneous catalysts because they are made of solid transition metals, like platinum or rhodium, while the reactants are gases – carbon monoxide (converted to carbon dioxide) and nitrogen dioxide (which ends up as nitrogen).

Biological systems like plants and animals contain catalysts called *enzymes*. Enzymes are all homogeneous catalysts.

Example 4

Figure 2.2 shows how the volume of gas made in a reaction changed with time.

Calculate the average rate of the reaction in the first 30 seconds.

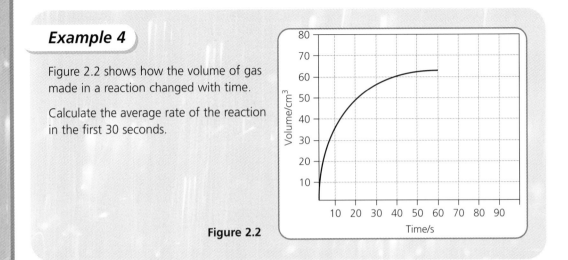

Figure 2.2

Solution

Remember how to calculate the average rate of a reaction?

$$\text{rate} = \frac{\text{change in concentration}}{\text{time taken}}$$

In this question, you can take the volume of gas as a measure of the concentration of product. In the first 30 seconds, the volume changes from zero up to 57 cm^3. The change is therefore 57 cm^3. Therefore

$$\text{rate} = \frac{57}{30} = \textbf{1.9 cm}^3\textbf{ s}^{-1}$$

Remember the units – in this case cm^3 was divided by seconds, so the units are $cm^3\,s^{-1}$.

The chances are that you'll be given the units in a question. It might be worded like this: 'Calculate the average rate of the reaction, in $cm^3\,s^{-1}$, in the first 30 seconds'. Or you might find a space for the answer with the units already given beside it. But if you're not given the units you need to put them in, or you could lose half a mark.

Example 5

Magnesium was made to react with dilute hydrochloric acid. In each experiment an excess of magnesium was used. Hydrogen gas is one of the products.

Which line in the table at the top of the following page correctly describes Experiment 2 when compared with Experiment 1?

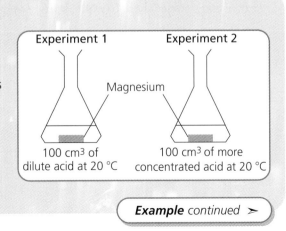

Example *continued* ➤

Example 5 continued

	Rate of reaction	Volume of hydrogen
A	Faster	More
B	Faster	Same
C	Slower	More
D	Slower	Same

Solution

Because the acid in Experiment 2 is more concentrated than the acid in Experiment 1, it's reasonable to assume that the reaction will be faster. So you have to pick from A or B.

That's the easy bit! Now, the magnesium is in excess in both experiments. This means that the amount of product will be controlled by the quantity of acid present. There's more acid present in 100 cm^3 of the more concentrated acid than there is in 100 cm^3 of the dilute acid, so you would expect to get more product – in this case, hydrogen. So **A** is the answer. There's more to this question than just knowledge and understanding of reaction rates – it involves a bit of problem solving as well!

Example 6

Catalytic converters speed up the conversion of harmful gases to less harmful gases. Which one of the following reactions is most likely to occur in a catalytic converter?

A   Carbon dioxide reacts to form carbon monoxide.

B   Carbon monoxide reacts to form carbon dioxide.

C   Nitrogen reacts to form nitrogen dioxide.

D   Oxygen reacts to form hydrogen oxide.

Solution

If you've taken in what you read in the summary section you should have no difficulty in choosing **B** as the answer!

The structure of the atom

Summary

Every element is built of *atoms*. The atoms themselves have a *nucleus* at the centre. The nucleus is built of *protons* – which have a *positive charge* and a mass of 1 atomic mass unit – and *neutrons* which are *neutral* and also have a mass of 1 atomic mass unit.

The nucleus is surrounded by *electrons*. Electrons have almost no mass and have a *negative charge*. Because atoms are neutral, there are the same numbers of protons and electrons. Because electrons have almost no mass, nearly all of the mass of an atom is contained in the nucleus. This is why the number of protons added to the number of neutrons is called the *mass number*. The number of protons is called the *atomic*

Figure 2.3 How we picture an atom

number and this fixes the identity of the element. For example, if an atom has 1 proton then it is an atom of hydrogen, but if it has 2 protons it is an atom of helium.

The identity of an element is not affected by the number of neutrons. Hydrogen atoms can have no neutrons or 1 neutron or 2 neutrons, but they are all still hydrogen atoms so long as they have just 1 proton. These are called *isotopes* of hydrogen. Isotopes of an element have the same atomic number (number of protons) but they have different mass numbers because they have different numbers of neutrons. Most elements are a mixture of isotopes. The average mass of these isotopes is called the *relative atomic mass*. For most elements, it's not a whole number because it is the average of several different mass numbers

Electrons

The electrons which surround the nucleus are arranged in *energy levels*. The lowest energy level is closest to the nucleus and can hold up to 2 electrons. Further out is a higher energy level which can hold up to 8 electrons. Even further out is a still higher energy level which can also hold up to 8 electrons. Beyond this are levels which hold electrons of even higher energy. Electrons always enter the lowest available energy level.

You'll find the electron arrangement for every element in the Data Booklet, but it's a great idea to be able to work out the electron arrangements for the first 20 elements without using the Data Booklet.

It's the number of outer electrons which fixes the way an atom behaves – so atoms with the same number of outer electrons have similar behaviour and are found in the same vertical column or group of the Periodic Table.

Summary continued ➤

Summary *continued*

Nuclide notation

The atomic number and mass number of an element are usually shown like this:

Mass number $\longrightarrow$
Atomic number $\longrightarrow$ $^{3}_{1}\text{H}$

The top number, 3, is the mass number – that is, the number of protons added to the number of neutrons. The bottom number, 1, is the atomic number – the number of protons. You can calculate the number of neutrons by subtracting the atomic number from the mass number. In this case, this isotope of hydrogen has 2 neutrons.

Example 7

An atom has atomic number 17 and mass number 35.

The number of neutrons in the atom is

A   17

B   18

C   35

D   52.

Solution

You ought to find this easy. Remember that the mass number is made up of the number of protons added to the number of neutrons. To get the number of neutrons, subtract the atomic number (the number of protons) from the mass number. Subtract 17 from 35 and get 18. This is the number of neutrons so the answer is **B**.

Remember – the atomic number is the number of protons, and the mass number is the number of protons added to the number of neutrons.

Example 8

Naturally occurring silver (atomic number 47, relative atomic mass 108) consists of a mixture of two isotopes with mass numbers 107(^{107}Ag) and 109 (^{109}Ag).

Identify the true statement.

A   Isotopes of silver have the same number of neutrons.

B   All silver atoms have a relative atomic mass of 108.

C   Atoms of ^{107}Ag are more abundant than those of ^{109}Ag.

D   All silver atoms have 47 electrons.

Solution

OK! This is a question about isotopes. You're expected to know that isotopes are atoms with the same atomic number but different mass numbers. That is, they have the **same** number of protons but **different** numbers of neutrons. That rules A out as an answer. You're also supposed to know that the relative atomic mass (RAM) is an **average** of the mass numbers of the isotopes. However, the RAM of ^{107}Ag is 107 and the RAM of ^{109}Ag is 109 so B isn't the answer either. If 108 is the average of the mass numbers of the isotopes (taking account of how much there is of each) then there must be equal amounts of ^{107}Ag and ^{109}Ag. This means that C is not the answer.

How about D? The atomic number of silver is 47. This tells us that it has 47 protons. It follows that there will be 47 electrons if the atom is neutral. So **D** is the answer.

Example 9

Complete the table for the particle shown.

Atomic number	Symbol for the element	Mass number

Key:
p = Proton
n = Neutron
• = Electron

Solution

Again, all you need to know is that the atomic number is the number of protons – in this case it's 10. The mass number is the combined number of protons and neutrons – that is 10 + 10 = 20. If you know the atomic number then you can identify the element using the Data Booklet – element number 10 is neon, with symbol Ne. So your table should look like this:

Atomic number	Symbol for the element	Mass number
10	Ne	20

Bonding, structure and properties

Summary

Bonding

Atoms are held together by *bonds*. *Covalent bonds* usually form between atoms of non-metal elements. They form when atoms share pairs of electrons – as shown in Figure 2.4.

The bond works because the positively charged nuclei of the atoms are attracted to the concentrated negative charge of the shared electrons which lie between the nuclei of the atoms. As a result of sharing electrons, each atom ends up with a *stable electron arrangement* which involves a full outer energy level – usually with 8 electrons, but 2 in the case of hydrogen.

Summary continued ➤

Summary *continued*

If the nuclei of the atoms which are sharing electrons have different attractions for electrons then the electrons will not be shared equally between the atoms. As a result, the bond becomes *polar*, and the molecule may end up with a positive end and a negative end.

Ionic bonds are forces of attraction between oppositely charged ions. Ions are atoms which have either gained or lost electrons. By doing this, they gain a stable electron arrangement involving a full outer energy level. Ionic bonds usually form between a metal element and a non-metal element.

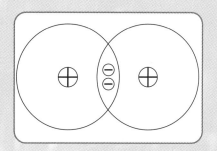

Figure 2.4 The two positive nuclei are attracted to the concentrated negative charge of the shared electrons

Metallic bonds hold metals together. The outer electrons of metal atoms are free to move throughout the metal (they are 'delocalised'). This leaves each metal atom with a positive charge. The attraction between the positive metal ions and the *delocalised electrons* holds the metal together.

Example 10

Metallic bonds are due to

A   pairs of electrons being shared equally between atoms

B   pairs of electrons being shared unequally between atoms

C   the attraction of oppositely charged ions for each other

D   the attraction of positively charged ions for delocalised electrons.

Solution
Well – if you've learned your work, this should be a really easy question. Answer A describes covalent bonds. Answer B describes polar covalent bonds. Answer C describes ionic bonds and **D**, the correct answer, describes metallic bonding. Easy, if you know your stuff!

Example 11

Hydrogen reacts with chlorine to form hydrogen chloride. Draw a diagram to show how the outer electrons are shared in a molecule of hydrogen chloride.

Solution

There's more than one way of showing this. You can show electrons as dots or crosses on rings which represent the energy levels or shells.

Chlorine is in Group 7, so it has 7 outer electrons.

You can show them like this.

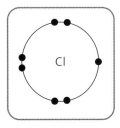

Don't forget to show that it's chlorine, by putting the symbol in the centre.

Hydrogen has just 1 electron in its outer shell. It can be shown like this. We'll use a cross to distinguish its electron from those of chlorine, and make the atom look smaller – because it is!

The diagram showing the outer electrons in hydrogen chloride looks like this.

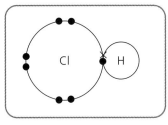

You might have been shown electrons in electron pair clouds rather than on a circle. Chlorine might look like this:

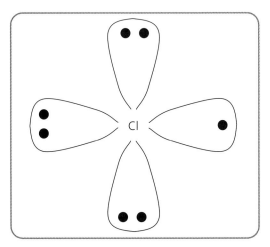

Hydrogen might look like this:

And hydrogen chloride might look like this:

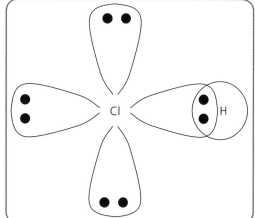

Whatever you do, make sure you show all the outer electrons on each atom, not just the electrons involved in forming the bond. And don't forget to put the symbols in, so you show which element is which.

Example 12

Draw a diagram to show how the outer electrons are shared in a molecule of nitrogen.

Solution
This is a bit of a challenge. If you show the nitrogen atoms like this:

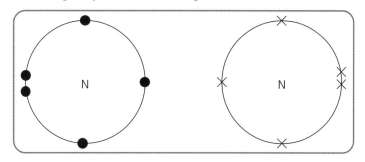

It's hard to see how you can show the overlap of three electrons from each atom. You just have to do it like this:

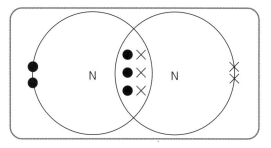

Or this:

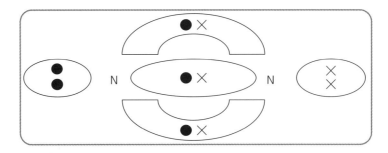

Example 13

The diagram represents the structure of a molecule of ammonia.

Why are the bonds between the nitrogen and the hydrogen in an ammonia molecule described as polar covalent?

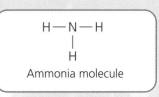

H—N—H
|
H
Ammonia molecule

Solution

 Watch out – are you being asked to explain the 'polar' or the 'covalent' or both?

In a covalent bond, the electrons are shared between atoms. If the bond is polar it's because the electrons are not shared equally. So you had better say that **the electrons are not shared between the atoms equally**. That gives the idea of sharing as well as explaining why the bond is polar. It would also be OK to say that the nitrogen attracts electrons more strongly than hydrogen.

Example 14

Which of the following pairs of elements combine to form an ionic compound?

A   Lead and fluorine

B   Sulphur and oxygen

C   Carbon and nitrogen

D   Phosphorus and oxygen

Solution

All you have to remember to solve this puzzle is that ionic compounds are usually formed between a metal and a non-metal. If you can spot a metal then you can answer the question. Lead is a metal – it's the only metal in the question. Fortunately, fluorine is a non-metal, so you pick **A** as the answer. The bonding is covalent in B, C and D as the elements involved are all non-metals.

Summary

Structure

When atoms are held together by covalent bonds, they form particles called *molecules*. A molecule built of two atoms is called *diatomic*. We saw earlier that the halogens, as well as hydrogen, oxygen and nitrogen, are diatomic. Remember – 'I bring clay for our new home'? The atoms don't have to be the same – carbon monoxide, CO, is also diatomic.

Figure 2.5 The Aurora Borealis (Northern Lights) are caused by solar particles hitting the diatomic molecules O_2 and N_2 in the atmosphere

You should be able to draw the *shapes of simple molecules*. Water, H_2O, is a bent molecule.

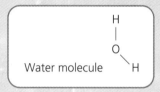

Water molecule

Some molecules – like ammonia, NH_3 – are *pyramidal*. The dotted line indicates a bond going backwards, and the wedge-shaped bond indicates a bond coming forwards.

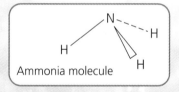

Ammonia molecule

Methane, CH_4, has a *tetrahedral* shape.

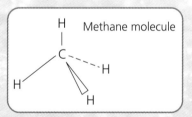

Methane molecule

You should practise drawing these shapes so you can do them with confidence when you meet them in the exam. As you can see from the examples above, the formula for a compound tells you how many atoms of each kind are contained in the molecule.

In a solid or liquid covalent substance, the molecules are held together by weak forces of attraction. Sometimes, instead of forming separate molecules, a substance exists as a *covalent network* in which all the atoms – billions of them – are all interconnected. Silicon dioxide, SiO_2, is like this. In this case, the formula tells you the ratio of the numbers of atoms

Summary continued ➤

Summary *continued*

– there are twice as many oxygen atoms as there are silicon atoms. Figure 2.6 shows a simplified diagram showing a tiny part of a silicon dioxide structure.

You have to remember that every oxygen is bonded to 2 silicons, and every silicon is bonded to 4 oxygens, so the structure goes on and on and on! (To make matters more complicated, it's actually a three-dimensional structure, not two as shown in the diagram).

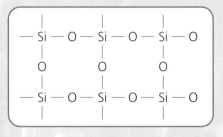

Figure 2.6 The covalent network structure of silicon dioxide

A *lattice* is formed when the bonding is ionic. In a lattice, each positive ion is surrounded by negative ions and each negative ion is surrounded by positive ions. The lattice is held together by the attraction of the opposite charges. Like the formula of a covalent network compound, the formula of an ionic compound tells you the *ratio* of the numbers of ions in the lattice. So sodium chloride, NaCl, contains 1 sodium ion for each chloride ion.

Figure 2.7 shows the idea – you need to remember that it continues in three dimensions.

Metals also have a lattice structure. You can think of this as a lattice of positive ions held together by the attraction of the delocalised electrons.

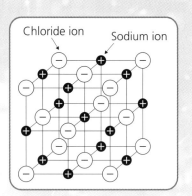

Figure 2.7 A typical ionic lattice – sodium chloride

Example 15

Carborundum is an example of a covalent network compound. The lattice contains equal numbers of silicon and carbon atoms:

Write the formula for carborundum.

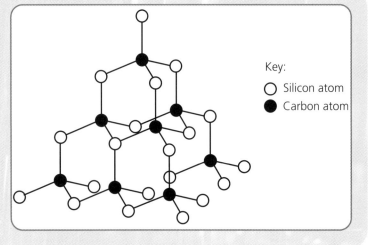

Key:
○ Silicon atom
● Carbon atom

Solution

Remember, in a covalent network compound the formula tells you the ratio of the number of atoms of each kind – it doesn't tell you the actual numbers of atoms. So if there are equal numbers of silicon and carbon atoms, the formula is just **SiC** – it's as simple as that!

Example 16

Which of the following compounds exists as diatomic molecules?

A   Carbon monoxide

B   Carbon tetrachloride

C   Nitrogen trihydride

D   Sulphur dioxide

Solution

To answer this question you need to remember what the prefixes mean:

◆   'mono' means 'one'

◆   'di' means 'two'

◆   'tri' means 'three'

◆   'tetra' means 'four'.

So a carbon monoxide molecule has 1 carbon atom and 1 oxygen atom – a total of 2 atoms. This means that carbon monoxide is diatomic – it has only two atoms in it. Carbon tetrachloride has 1 carbon and 4 chlorine atoms – 5 altogether. Nitrogen trihydride has 1 nitrogen atom and 3 hydrogen atoms – 4 altogether. Sulphur dioxide has 1 sulphur atom and 2 oxygen atoms – 3 altogether. So the answer is **A**. Of course, you shouldn't have to work it out like this – you should just learn that carbon monoxide is diatomic!

Example 17

In ammonia molecules, the atoms are held together by three covalent bonds.

a)   What is a covalent bond?

b)   The formula of ammonia is NH_3. Draw a diagram to show the shape of an ammonia molecule.

Solution

a)   It might be enough to say that a covalent bond forms when **two atoms share a pair of electrons**. It's probably better to add that this gives each atom a stable electron arrangement.

b)   You're expected to know that the ammonia molecule has a **pyramidal** structure, and if you've been sensible you'll have practised drawing it. This is what you should draw.

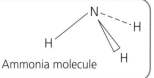

Ammonia molecule

23

HOW TO PASS INTERMEDIATE 2 CHEMISTRY

Summary

Properties

The properties of a substance result from the bonding in it. Metallic bonding is strong, and involves delocalised electrons. As a result, metals (except mercury) are solids with quite high melting points and all conduct electricity. Ionic bonding is strong, so ionic compounds are also solids with high melting points. They don't conduct electricity when solid because the ions can't move. When they are melted, the ions can move and the melt conducts electricity. If they are dissolved in water, the ions can move so the solution conducts electricity. Many ionic compounds dissolve in water.

Remember that an electric current is a flow of charged particles – these can be electrons as in metals, or ions as in ionic compounds.

Covalent compounds have no charged particles and only weak bonds between the molecules. So covalent substances tend to be gases, liquids or solids with low melting points. They don't conduct electricity under any circumstances. If it's a covalent network then it will have a very high melting point because all the atoms are covalently bonded to each other. Many covalent compounds are insoluble in water, but they may dissolve in other solvents.

If a direct current (d.c.) passes through a molten or dissolved ionic compound, chemical changes take place at the electrodes. The negative electrode supplies electrons to the positive ions, giving them a stable electron arrangement. They turn into metal atoms (because positive ions are usually metal ions) or hydrogen (if the solution contains lots of positive hydrogen ions). The positive electrode removes electrons from negative ions, giving them a stable electron arrangement and forming molecules of non-metal elements, because non-metals usually form negative ions.

Example 18

Which one of the following elements conducts electricity?

A   Bromine

B   Mercury

C   Oxygen

D   Sulphur

Solution

All you need to know here is that all metals conduct electricity as a result of their delocalised electrons. OK – spot the metal. You could always check each element out in the Periodic Table and find one to the left of the famous zig-zag line. But you should know that mercury is a metal so that **B** is the answer.

Example 19

Why does copper conduct electricity?

Solution
The answer is the same as the first sentence of the last solution – **it has delocalised electrons that are free to move**.

Chemical symbolism

Summary

Formulae

You've got to be confident at writing chemical formulae. If it's a simple, two-element compound then you can check what groups the elements are in from the Data Booklet. The combining power of an atom is the same as the number of unpaired electrons it has.

Group	1	2	3	4	5	6	7
Combining power (valency)	1	2	3	4	3	2	1

Suppose you want the formula for calcium chloride. Calcium is in Group 2 so its combining power is 2. Chlorine is in Group 7 so its combining power is 1. The combining powers of each element have to be the same – so you need 2 chlorines for each calcium. The formula is written $CaCl_2$. You can get the formula quickly by swapping the combining powers.

Aluminium oxide's formula can be found this way. Aluminium has a combining power of 3 and oxygen has a combining power of 2. If you swap these numbers over you get Al_2O_3.

If a Roman numeral is used in the name of a compound, that's just the combining power of the element it refers to. For example, iron(III) oxide tells you that the combining power of the iron is 3, so the formula would be Fe_2O_3. Roman numerals are often used when an element can have more than one valency.

Watch out! If you're asked to work out a formula like that of silicon oxide – it looks as if it's Si_2O_4 but this simplifies to SiO_2.

If you have to write the formula of a compound involving an ion containing more than one element then use the Data Booklet to be sure of the formula of the ion. Its combining power is the same as its charge – so, for example, the combining power of the sulphate ion, SO_4^{2-}, is 2.

Don't forget to use brackets if the formula contains more than one of these ions. For example, the formula of calcium hydroxide is written $Ca(OH)_2$.

Example 20

Write the formula for

a) magnesium iodide
b) aluminium sulphide
c) phosphorus oxide
d) iron(II) chloride
e) tin(IV) oxide
f) sodium sulphate
g) ammonium sulphide
h) ammonium sulphate
i) copper(II) hydroxide
j) iron(III) sulphate

Solution

a) MgI_2
b) Al_2S_3
c) P_2O_3
d) $FeCl_2$
e) SnO_2 (don't forget to simplify!)
f) Na_2SO_4
g) $(NH_4)_2S$
h) $(NH_4)_2SO_4$
i) $Cu(OH)_2$
j) $Fe_2(SO_4)_3$

Summary

Equations

Questions about equations often give you the equation and ask you to balance it. You can also be asked to write ion–electron equations, but we'll deal with this in Unit 3.

Example 21

The equation refers to reactions which take place in a catalytic converter:

$$CO + NO \rightarrow CO_2 + N_2$$

Balance this equation.

Solution

It appears that the carbon and oxygen are already balanced. However, there's only 1 nitrogen on the left-hand side and 2 on the right-hand side. So you'll need 2NO on the left:

$$CO + 2NO \rightarrow CO_2 + N_2$$

Now there isn't enough oxygen on the right! You need $2CO_2$ on the right:

$$CO + 2NO \rightarrow 2CO_2 + N_2$$

Now the left is short of 1 C and 1 O so you need to add another CO to the left:

$$2CO + 2NO \rightarrow 2CO_2 + N_2$$

It's done! You might have spotted that this could have been done a lot faster if you had just written:

$$CO + NO \rightarrow CO_2 + \tfrac{1}{2}N_2$$

Some students don't like using '½' in an equation, but it's perfectly acceptable to do this.

Example 22

A new type of air bag is being developed for use in cars. In the reaction, butane reacts with an oxide of nitrogen:

$$C_4H_{10} + N_2O \rightarrow CO_2 + H_2O + N_2$$

Balance this equation.

Solution

Note that there are 4 carbons on the left-hand side and only 1 on the right-hand side. You need 4 carbon dioxides on the right to balance this:

$$C_4H_{10} + N_2O \rightarrow 4CO_2 + H_2O + N_2$$

Now there are 10 hydrogens on the left and only 2 on the right. You need 5 water molecules to balance this:

$$C_4H_{10} + N_2O \rightarrow 4CO_2 + 5H_2O + N_2$$

On the left now there is only 1 oxygen but there are 13 on the right – 8 in the carbon dioxide and 5 in the water. You need 13 nitrogen oxides on the left to give 13 oxygens:

$$C_4H_{10} + 13N_2O \rightarrow 4CO_2 + 5H_2O + N_2$$

Finally, 26 nitrogens are needed on the right to complete the balance:

$$C_4H_{10} + 13N_2O \rightarrow 4CO_2 + 5H_2O + 13N_2$$

The mole

Summary

Chemists measure the amount of a substance in *moles*. To work out the mass of 1 mole of substance, you *add up the relative atomic masses* (RAM) of each element present.

Figure 2.8 Not a chemical mole

Summary continued ➤

Summary *continued*

Don't forget to multiply by the number of atoms present. For example, if you want to know the mass of a mole of sodium chloride, NaCl, go to page 4 of the Data Booklet and add together 23 (the RAM of sodium) and 35.5 (the RAM of chlorine) giving 58.5. A mole of sodium chloride weighs 58.5 g.

If it's calcium chloride, $CaCl_2$, we add 40 and (2×35.5) to get 111. A mole of calcium chloride weighs 111 g. If the formula involves brackets, you multiply the mass of the elements inside the brackets by the number outside the bracket.

If you're told the mass of a substance and want to calculate the number of moles present, you divide the given mass by the mass of 1 mole:

$$\text{number of moles} = \frac{\text{given mass}}{\text{mass of 1 mole}}$$

Example 23

What is the relative formula mass of ammonium sulphate, $(NH_4)_2SO_4$?

A   70

B   118

C   132

D   228

Solution

Add up the elements inside the brackets first – there's 1 nitrogen and 4 hydrogens – that's $14 + 4 = 18$. You have to double this because there's a 2 outside the bracket, making 36. To this you must add 32 for the sulphur, and $4 \times 16 = 64$ for the four oxygens. This all comes to 132. Hooray – it's one of the given answers so it must be right. **C** is the answer.

Example 24

Urea has the formula H_2NCONH_2. An adult male, on average, excretes 30 g of urea each day in urine. How many moles are there in 30 g of urea?

Solution

The first thing you need to do is calculate the mass of 1 mole of urea. The numbers you add together are:

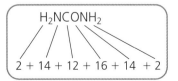

This comes to 60, so 1 mole of urea weighs 60 g. This question involves 30 g, so you calculate the number of moles by dividing 30 g by 60 g (remember – divide the given mass by the mass of 1 mole):

$$\frac{30}{60} = 0.5$$

The answer is **0.5 moles**.

Example 25

How many moles are in:

a)   14 g of KOH

b)   5 g of $CaCO_3$

c)   53 g of Na_2CO_3

d)   5 g of NaOH

e)   29.25 g of NaCl

f)   14.8 g of $Ca(OH)_2$

g)   8.2 g of $Ca(NO_3)_2$

h)   2.8 g of CaO

Solution

a)   **0.25 moles**

b)   **0.05 moles**

c)   **0.5 moles**

d)   **0.125 moles**

e)   **0.5 moles**

f)   **0.2 moles**

g)   **0.05 moles**

h)   **0.05 moles**

Reacting amounts

Summary

You can use balanced equations and the idea of the mole to work out how much of a substance is used up or formed in a reaction.

Example 26

When dinitrogen monoxide decomposes it forms a mixture of nitrogen and oxygen:

$$2N_2O \rightarrow 2N_2 + O_2$$

How many moles of oxygen will be produced when 4 moles of dinitrogen monoxide are decomposed?

Solution

This equation tells you that 2 moles of dinitrogen monoxide decompose to give 2 moles of nitrogen and 1 mole of oxygen:

$$2N_2O \rightarrow 2N_2 + O_2$$

2 moles   2 moles   1 mole

Therefore, if you start with 4 moles of dinitrogen monoxide you can expect **2 moles** of oxygen.

Example 27

The equation for the preparation of magnesium sulphate from magnesium is shown below:

$$Mg + H_2SO_4 \rightarrow MgSO_4 + H_2$$

Calculate the mass of magnesium sulphate produced when 4.9 g of magnesium reacts completely with sulphuric acid.

Solution

Note that the question is about magnesium and magnesium sulphate. Decide how many moles of each are involved in the equation and write it down:

$$Mg + H_2SO_4 \rightarrow MgSO_4 + H_2$$

1 mole   1 mole

Then work out what they weigh:

$$Mg + H_2SO_4 \rightarrow MgSO_4 + H_2$$

1 mole   1 mole

24.5 g   120.5 g

Scale the numbers down to 1 g of magnesium by dividing all numbers by 24.5:

$$Mg + H_2SO_4 \rightarrow MgSO_4 + H_2$$

1 mole   1 mole

24.5 g   120.5 g

1 g   4.918 g

Scale the numbers up to 4.9 g of magnesium by multiplying them all by 4.9:

$$Mg + H_2SO_4 \rightarrow MgSO_4 + H_2$$

1 mole   1 mole

24.5 g   120.5 g

1 g   4.918 g

4.9 g   24.1 g

So the answer is **24.1 g** of magnesium sulphate.

Example 28

Ammonia gas can be produced in the lab by heating ammonium chloride with sodium hydroxide:

$$NH_4Cl + NaOH \rightarrow NaCl + H_2O + NH_3$$

Calculate the mass of ammonia produced by reacting 10 g of ammonium chloride with excess sodium hydroxide.

Solution

Note that the question is about ammonia and ammonium chloride. Decide how many moles of each are involved in the equation and write it down:

$$NH_4Cl + NaOH \rightarrow NaCl + H_2O + NH_3$$

1 mole   1 mole

Then work out what they weigh:

$$NH_4Cl + NaOH \rightarrow NaCl + H_2O + NH_3$$

1 mole   1 mole

53.5 g   17 g

Scale the numbers down to 1 g of ammonium chloride by dividing all numbers by 53.5:

$$NH_4Cl + NaOH \rightarrow NaCl + H_2O + NH_3$$

1 mole   1 mole

53.5 g   17 g

1 g   0.318 g

Scale the numbers up to 10 g of ammonium chloride by multiplying them all by 10:

$$NH_4Cl + NaOH \rightarrow NaCl + H_2O + NH_3$$

1 mole   1 mole

53.5 g   17 g

1 g   0.318 g

10 g   3.18 g

So the answer is **3.18 g** of ammonia.

Here are more examples for you to try. First you have to balance the equation (if necessary) and then carry out a calculation based on the equation.

Example 29

a) What mass of sulphur trioxide is formed from the complete reaction of 8 g of sulphur dioxide with oxygen?

$$SO_2 + O_2 \rightarrow SO_3$$

b) What mass of HCl would be formed from the complete reaction of 10 g of H_2 with chlorine?

$$H_2 + Cl_2 \rightarrow HCl$$

c) What mass of carbon dioxide would form from the complete combustion of 7 g of C_2H_4?

$$C_2H_4 + O_2 \rightarrow CO_2 + H_2O$$

d) What mass of nitrogen would form if 34 g of ammonia (NH_3) reacted completely to give nitrogen and water?

$$NH_3 + O_2 \rightarrow N_2 + H_2O$$

e) What mass of hydrogen does 1.4 g of nitrogen require for a complete reaction to form ammonia?

$$N_2 + H_2 \rightarrow NH_3$$

f) What mass of oxygen does 5 g of hydrogen require for complete combustion to form water?

$$H_2 + O_2 \rightarrow H_2O$$

g) What mass of nitrogen is required to form 23 g of nitrogen dioxide, NO_2?

$$N_2 + O_2 \rightarrow NO_2$$

h) What mass of carbon dioxide, CO_2, will result from the complete combustion of 4.4 g of propane, C_3H_8?

$$C_3H_8 + O_2 \rightarrow CO_2 + H_2O$$

The balanced equations:

a) $2SO_2 + O_2 \rightarrow 2SO_3$

b) $H_2 + Cl_2 \rightarrow 2HCl$

c) $C_2H_4 + 3O_2 \rightarrow 2CO_2 + 2H_2O$

d) $2NH_3 + 1\tfrac{1}{2}O_2 \rightarrow N_2 + 3H_2O$

e) $N_2 + 3H_2 \rightarrow 2NH_3$

f) $2H_2 + O_2 \rightarrow 2H_2O$

g) $N_2 + 2O_2 \rightarrow 2NO_2$

h) $C_3H_8 + 5O_2 \rightarrow 3CO_2 + 4H_2O$

The answers to the calculations:

a) **10 g**

b) **365 g**

c) **22 g**

d) **28 g**

e) **0.3 g**

f) **40 g**

g) **7 g**

h) **13.2 g**

Chapter 3

CARBON COMPOUNDS

In this unit you find out about all the different kinds of carbon compounds and how they behave. You'll find that carbon compounds include such important substances as fuels (almost every fuel contains carbon), plastics and foods.

There are a lot of new words in this unit, so you'll have to make a special effort to remember them!

The unit covers the following topics.

◆ Fuels

◆ Nomenclature and structural formulae

◆ Reactions of carbon compounds

◆ Plastics and synthetic fibres

◆ Natural products

Fuels

Summary

A *fuel* is a substance that reacts with oxygen to produce *energy* in an exothermic reaction. The reaction involves burning the fuel – this is called *combustion*. Many fuels are *hydrocarbons*, which contain the elements carbon and hydrogen only. When we burn these fuels, the carbon combines with oxygen to form carbon dioxide and the hydrogen combines with oxygen to form water. Carbon dioxide turns lime water cloudy – this is the usual test for carbon dioxide. If there's not enough oxygen present when the fuel is burning then carbon monoxide, a poisonous gas, forms instead. Solid carbon can also form – that's what soot is.

Vehicles with diesel engines can produce soot particles, which are very harmful if inhaled. Car engines burn hydrocarbon fuels. This means that they produce carbon dioxide and carbon monoxide. They can also produce oxides of nitrogen as the result of electric sparks passing through the mixture of fuel and air in the car engine. However, *catalytic converters* turn carbon monoxide and nitrogen oxides into carbon dioxide and nitrogen – less dangerous gases.

Some fuels, like coal, contain sulphur as an impurity. When the coal burns, poisonous sulphur dioxide escapes into the air.

Summary *continued* ➤

33

Summary continued

Many fuels are made from *crude oil*, which is a mixture of hydrocarbons found in the Earth's crust. These are separated from each other in oil refineries in a process called *fractional distillation*, in which separation into *fractions* is carried out. This is possible because different hydrocarbons have different boiling points. A fraction is just a set of hydrocarbons with boiling points between a lower limit and an upper limit.

The smaller the hydrocarbon molecule, the lower the boiling point. Hydrocarbons with low boiling points turn easily into vapour. This makes the hydrocarbon more flammable. Also, the smaller the molecules, the thinner (less viscous) the hydrocarbon is.

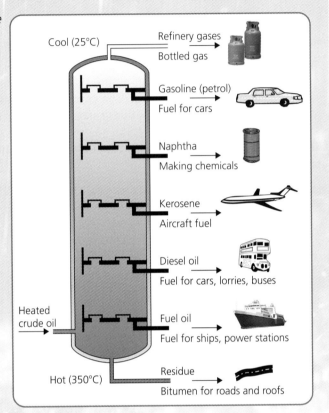

Figure 3.1 Fractional distillation is carried out in a refinery tower

Example 1

The fractional distillation of crude oil depends on the fact that different hydrocarbons have different

A densities

B solubilities

C boiling points

D ignition temperatures.

Solution

Just a simple recall of fact – the process depends on different boiling points so the answer is **C**.

Example 2

Fuel oil (see Figure 3.1) produced by the fractional distillation of crude oil has a high viscosity.

Which of the following properties also apply to fuel oil?

A   Low boiling point and high flammability

B   High boiling point and high flammability

C   Low boiling point and low flammability

D   High boiling point and low flammability

Solution

Because fuel oil is very viscous, it must be made of very large molecules – it's viscous because the molecules all get tangled up with each other (Just like spaghetti!). Because the molecules are big, it will be difficult for them to turn into vapour – that is, the boiling point will be high. A liquid must turn into vapour before it can burn, so it will therefore not be very flammable. So the answer is **D**.

Figure 3.2 Long hydrocarbon molecules get tangled up like spaghetti

Nomenclature and structural formulae

Summary

Hydrocarbons

You're expected to know about three families of hydrocarbons:

◆   the *alkanes* – contain only carbon to carbon single bonds;

◆   the *alkenes* – contain at least one carbon to carbon double bond;

◆   the *cycloalkanes* – form rings of carbon atoms.

You'll find their names, in order of increasing numbers of carbon atoms, on page 6 of the Data Booklet. You should be able to draw any of these structures confidently, and also to be able to recognise a structure and name it. You could make yourself a set of flash cards for this.

You should also be able to write shortened structural formulae. For example, butane is:

and its shortened structural formula is $CH_3CH_2CH_2CH_3$. Its molecular formula is C_4H_{10}.

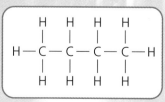

Summary continued ➤

Summary *continued*

When there are branches on the alkane molecule, you follow simple rules to name it.

1   Find the longest chain of carbon atoms – this gives you the name of the parent alkane. Don't be deceived by bends or kinks in the chain.

2   Branches are named from the number of carbon atoms they contain. The -ane ending of the alkane is changed to -yl, For example, methane becomes 'methyl'; ethane becomes 'ethyl'.

3   If there are 2 methyl groups, we say 'dimethyl'.

4   Number the chain of carbons from the end which gives the smallest number for the first branch.

You'll also find the alkenes on page 6 of the Data Booklet. When naming them, don't forget to put in the position of the double bond:

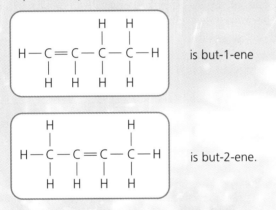

is but-1-ene

is but-2-ene.

Make sure that you can draw the structural formulae of the cycloalkanes – again, the names are given in the Data Booklet in order of increasing numbers of carbon atoms, so you shouldn't put the wrong number of carbon atoms in the ring.

Don't forget the general formulae for these sets of compounds. The alkanes follow the formula C_nH_{2n+2}. The alkenes and cycloalkanes follow the formula C_nH_{2n}.

Make sure you know the definition of a *homologous series* – a set of compounds with the same general formula and similar chemical properties. And make sure you know that *isomers* are molecules with the same molecular formula but different structural formulae. Don't mix this up with isotopes. Learn which is which – 'isomers' has an '**m**' in the middle so use that to remind you of the 'same **m**olecular formula'; 'isotopes' has a '**p**' in the middle so use that to remind you of the 'same number of **p**rotons'.

Example 3

Which one of the following molecules is an isomer of heptane?

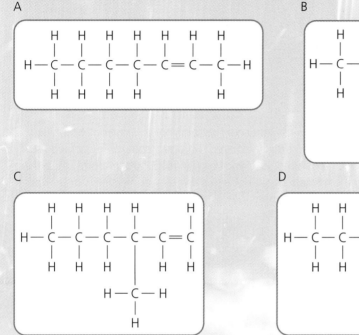

A

B

C

D

Solution

You're looking for an isomer of heptane. It's really important that you read the name carefully – not just the start of the name or just the end of the name – the whole name. It's heptane. Check out the Data Booklet and you'll see that 'hept-' means 7 carbon atoms in the molecule. So check them all out. A has 7 – a possibility? B has 6. C has 7 – another possibility. D also has 7. However, the name you want ends in '-ane' so no double bonds are allowed, otherwise the name would end in '-ene'. That rules out A and C, so you're left with D. It's got the right number of carbon atoms and no double bonds. **D** is the answer.

Example 4

Which one of the following compounds has isomers?

A

```
      H   H
      |   |
  H — C — C — H
      |   |
      H   H
```

B

```
  H       H
  |       |
  C  =  C
  |       |
  H       H
```

Example continued ➤

Example 4 continued

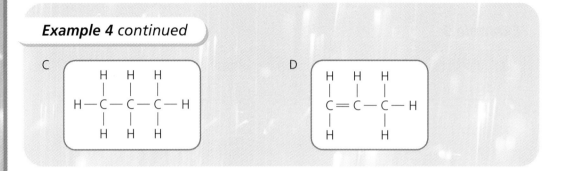

Solution

To solve this question, you need to think. You could certainly try to draw out possible isomers of each molecule, but this isn't really required. Different isomers are formed by different arrangements of the carbon atoms. If you've only got 2 carbon atoms, then there's only one way of arranging them – side by side! There aren't any isomers of A or B – there aren't enough carbon atoms. There's no other way of arranging 3 carbon atoms when they're linked by single bonds. However, the presence of a double bond in D gives it a molecular formula of C_3H_6. Remember, C_nH_{2n} is the general formula for the alkenes and the cycloalkanes. Cyclopropane is an isomer of this propene molecule. So the answer is **D**. In general, the more carbon atoms a molecule has, the more likely it is to form isomers.

Example 5

The name of this compound is

A   2,3-dimethylpentane

B   3,4-dimethylpentane

C   2,3-dimethyl propane

D   3,4-dimethylpropane.

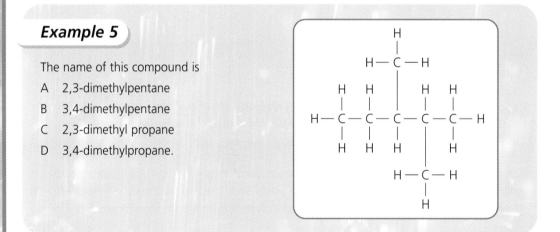

Solution

The longest continuous chain of carbon atoms is a chain of 5. Therefore, this compound is based on pentane. Not propane!! Remember, the names are given in the Data Booklet in order of increasing numbers of carbon atoms, so there's no excuse for picking C or D. The answer is A or B. Remember that the carbon atoms are numbered so that the first branch has the lowest possible number. That means that you'll have to count from the right end of this molecule, giving 2,3-dimethylpentane. **A** is the answer.

Example 6

Two isomers of butene are:

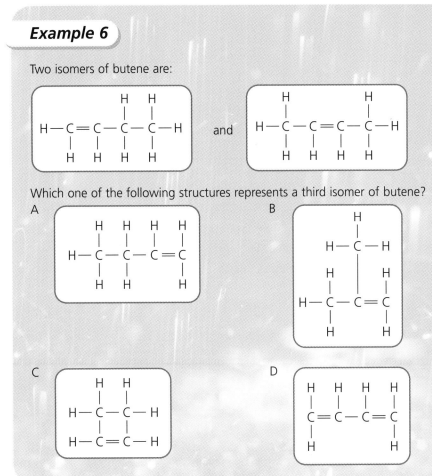

Which one of the following structures represents a third isomer of butene?

Solution

To start off with, remember what isomers are. They're molecules with the same molecular formula and different structural formulae. They'll have the same molecular formula if they have the same number of carbon atoms and hydrogen atoms. The molecules in this question both have the formula C_4H_8.

A and B both have this molecular formula. C and D both have the formula C_4H_6. Neither C nor D can be the answer. Are there two correct answers – A and B? Look closely at A. Try to see that it's the same as the first molecule in the question, just flipped over. (In fact, they're both but-1-ene). If it's the same then it can't be an isomer. This leaves B. It has the right molecular formula, but its structure is clearly different because it's got a branch. In fact, its name is 2-methylpropene. If it has a different name then it has a different structure – it's also different from the other molecule at the start, but-2-ene. So the answer is **B**.

Example 7

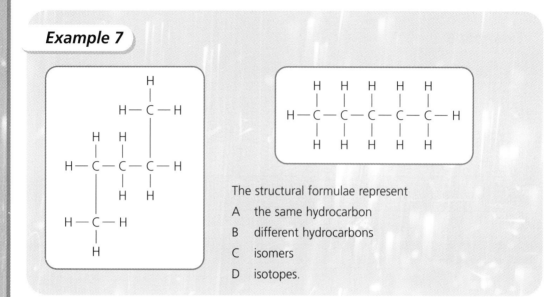

The structural formulae represent

A   the same hydrocarbon

B   different hydrocarbons

C   isomers

D   isotopes.

Solution

OK – D can be ruled out because isotopes are atoms with the same number of protons and different numbers of neutrons, and have nothing to do with this question. Remember that you were told earlier to make sure that you know the difference between isotopes and isomers.

So are they isomers? Count up the carbon atoms and the hydrogen atoms – both are C_5H_{12} so they have the same molecular formula. To be isomers, they have to have different structural formulae. Well, they look different – but if you think about it they both have a chain of 5 carbon atoms one after the other. They're actually the same – putting bends into a chain doesn't make them different hydrocarbons. The answer is **A**.

Summary

Alkanols, alkanoic acids and esters

You can think of **alkanols** as being alkanes which have had an H atom replaced by a hydroxyl group, – OH.

So CH_3OH is methanol and C_2H_5OH is ethanol. With propanol, there are two possibilities:

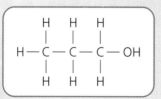
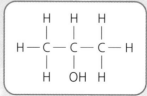

Summary *continued* ➤

Summary *continued*

In the first molecule, the hydroxyl group, – OH, is at the end. It's propan-1-ol. In the second molecule the hydroxyl group is second from the end, so it's propan-2-ol.

If you forget how to name alkanols you'll find a reminder on page 6 of the Data Booklet. Remember, unless it's methanol or ethanol you've got to put a number in the name to show where the hydroxyl group is.

Alkanoic acids contain the carboxyl group, which is also written as – COOH. Methanoic acid is HCOOH, ethanoic acid is CH₃COOH and so on.

When you're drawing the structural formula of an alkanol or an alkanoic acid, be careful with the – OH group. Written out fully it would be:

$$-O-H$$

showing that oxygen forms 2 bonds and hydrogen forms only 1 bond.

! **If you draw a carboxyl group like this, you're in danger of losing half a mark because the bond isn't pointing straight at the oxygen. Take care!**

Suppose you're asked to draw the **full** structural formula for ethanoic acid – this is what you have to draw:

You might even get away with this:

although it's not a **full** structural formula – and neither is:

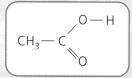

Summary *continued* ➢

Summary *continued*

If a question says '**Draw a** structural formula for ...' instead of '**Draw the full** structural formula for ...' then you can get away with grouping atoms together, like – OH instead of – O – H or with – CH_3 instead of showing all the bonds in a methyl group. A full structural formula shows all the bonds. However, if your structural formula contains enough detail to convince the marker that you know the structure, you'll probably get the mark.

Alkanols and alkanoic acids can combine to form **esters**.

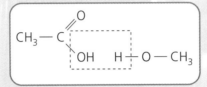

The 2 Hs and the O are removed as water, and the remaining bits join up to form an ester.

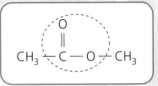

Figure 3.3 The ester link

The ester shown in Figure 3.3 is methyl ethanoate – the first bit of the name comes from the alkanol used in the reaction and the second bit of the name comes from the alkanoic acid.

Reactions in which small molecules, like water, are pulled out, while the remaining bits of molecules join up, are called *condensation* reactions. The atoms in the dotted oval form an *ester link*.

Figure 3.4 Raspberries contain at least 27 different esters

Example 8

Which one of the following structural formulae is that of an ester?

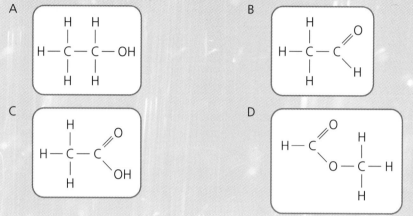

A

B

C

D

Solution

You're looking for an ester, so you should look out for an ester link as shown in Figure 3.3 in the dotted oval.

You should recognise A as an alkanol, ethanol, because of the presence of the hydroxyl group. You won't recognise B (until you do Higher Chemistry) but it's certainly not an ester – there's no ester link. You should recognise C as an alkanoic acid (ethanoic acid) because of the carboxyl group, so you're left with **D** which has indeed got an ester link.

Example 9

The structural formulae for two of the compounds in lavender oil are shown here.

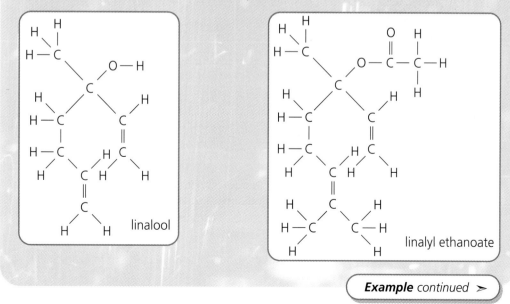

linalool

linalyl ethanoate

Example continued ➤

43

Example 9 continued

a) Linalool is an alkanol. Circle the alkanol group in the linalool molecule shown above.

b) Linalyl ethanoate is made from an alkanol and ethanoic acid. To which group of compounds does linalyl ethanoate belong?

c) Draw the full structural formula for ethanoic acid.

Solution

a) You're supposed to know that the group of atoms that makes a compound an alkanol is the hydroxyl group, $-OH$. So that's what you have to circle!

b) If it's made from an alkanol and ethanoic acid, which is an alkanoic acid, then it must be **an ester**. Besides, there's an ester link on the molecule – it's in the top right-hand corner of the molecule.

c) You're asked to draw the **full** structural formula for ethanoic acid so this is what you have to draw:

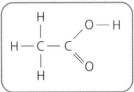

Remember – in a full structural formula, show all the bonds!

Example 10

Which one of the following is a structural formula for methyl ethanoate?

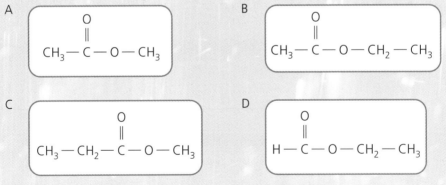

Solution

If the ester is called methyl ethanoate, then it's built of methanol (remember, the first bit of the ester name comes from the alkanol) and ethanoic acid. Methanol has just 1 carbon atom, and ethanoic acid has 2. This gives a total of 3 carbon atoms, so the ester has to contain 3 carbon atoms. This rules out B and C, because they both have 4 carbon atoms.

That leaves A and D. Remember the structural formula of ethanoic acid from Example 9?

Notice that the acid has a C = O group in it. That will help us to spot which of A and D is made from ethanoic acid. In A, the C = O identifies the acid present as having 2 carbon atoms – ethanoic acid. So **A** must be the answer. The alkanol bit of the ester is at the right-hand side, with 1 carbon atom – methanol.

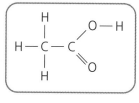

D isn't the answer because the acid has 1 carbon atom (so it's methanoic acid) and the alkanol has 2 carbon atoms (so it's ethanol). This ester is ethyl methanoate – and that's not methyl ethanoate – you need to read questions carefully.

Reactions of carbon compounds

Summary

Addition

Alkenes can add atoms across their double bond. Bromine is the most common example but the other halogens, as well as hydrogen and water, can also be added. The double bond ends up as a single bond. From being *unsaturated* the compound becomes *saturated*. The addition of bromine is used as a test for unsaturation – alkenes decolourise it quickly.

Cracking

This is done to long hydrocarbon molecules to break them into shorter ones, for which there is a greater demand. Heat is used along with a catalyst to reduce the amount of energy required. Cracking always results in a mixture of alkanes and alkenes because there are not enough hydrogen atoms present to form only alkanes.

Making and breaking esters

We saw earlier that esters are formed from alkanols and alkanoic acids in a condensation reaction. Concentrated sulphuric acid is used as a catalyst, along with a warm water bath as a source of heat. The esters that you will have met have pleasant fruity smells. You can break an ester down into its original alkanol and alkanoic acid by treating it with acids or alkalis. This process is called *hydrolysis*.

Reactions involving ethanol

Ethanol forms, along with carbon dioxide, when solutions of glucose undergo *fermentation* in the presence of *enzymes* in yeast. The process usually stops when the ethanol content reaches about 14 or 15% – high enough to kill the yeast, but it depends on the kind of yeast used. The ethanol solution can be made more concentrated by distillation.

Ethanol can also be made by adding water to ethene. The double bond opens and a hydrogen atom bonds to one carbon and an – OH group bonds to the other. This can be called *hydration* as well as addition. The reaction can be reversed and ethanol can be dehydrated to ethene.

Ethanol provides a *renewable* source of energy as a fuel because it is flammable and can be made from renewable crops such as sugar cane.

Example 11

Ethanol vapour can be dehydrated by passing it over hot aluminium oxide. Which one of the following compounds would be produced?

A   Ethane

B   Ethene

C   Ethanoic acid

D   Ethyl ethanoate

Solution

There really is not much to this question. You're told it's a dehydration reaction and you should have learned that ethanol can be dehydrated to ethene (which, of course, can be hydrated to ethanol). If you're in doubt, think about the formula for ethanol, C_2H_5OH. Think about removing H_2O – 2 hydrogens and 1 oxygen – and you're left with C_2H_4, the formula for ethene. The answer is **B**.

Example 12

Ethyl ethanoate can be made by reacting ethanoic acid and ethanol:

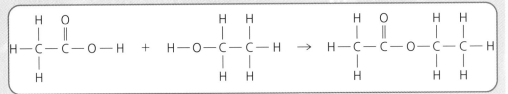

a)   Name this type of chemical reaction.

b)   Another method of making ethyl ethanoate from ethanol only has been developed.

$$2C_2H_5OH \quad \xrightarrow{\text{catalyst}} \quad CH_3COOC_2H_5 + 2X$$

Name substance X.

c)   Another method was developed for use in countries where ethanol is made from a renewable source. Name this source of ethanol.

Solution

a)   Well, this doesn't pose much of a challenge! All you need is to know that when esters form, it's a **condensation reaction**. (And you should be just as sure that when esters are broken down, it's a hydrolysis reaction!)

b)   This is really problem solving (there's a chapter on this later on). If you count up the atoms on the left and right of the arrow, you'll see that there are 4 carbons on each side. X doesn't therefore contain carbon. Check out the hydrogens – there are 12 on the left but only 8 on the right. The 2X must contain 4 hydrogens. There are 2 oxygens on the left and 2 on the right. So 2X must be $2H_2$ so X is H_2. Note that the question says 'Name substance X' so you should write the word '**hydrogen**' as the answer.

If you give a formula where the question says 'name', you will lose the mark if the formula is wrong – but not if you spell the name wrongly!

c)   Learn your stuff – learn that it's **sugar cane**.

Example 13

Propene can take part in addition reactions:

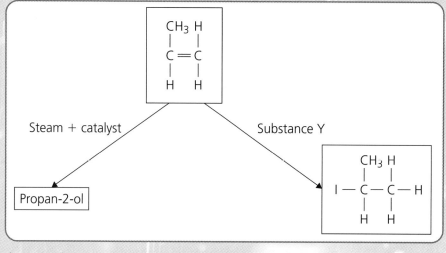

a)   Draw the structural formula for propan-2-ol.

b)   Identify substance Y.

Solution

a)   There should be no difficulty in drawing the structural formula for propan-2-ol. You need 3 carbon atoms, with a hydroxyl group on the second (middle) carbon atom. Your structure should look like this:

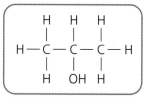

When you're drawing a structural formula, always make sure that every carbon atom forms 4 bonds – a double bond counts as 2, of course. This isn't a full structural formula because it doesn't show the bond between the oxygen and the hydrogen – but the question only asks for 'the structural formula' – not 'the full structural formula'.

b)   You should notice that the double bond in the propene 'disappears'. At the same time, there is an iodine atom at one end of the molecule and an extra hydrogen at the other end. **Hydrogen iodide**, HI, has been added so that must be what substance Y is.

Plastics and synthetic fibres

Summary

Addition polymerisation

Polymers are made up of long molecules formed by the combination of large numbers of small molecules called *monomers*. *Addition polymers* are made from monomers which have a carbon to carbon double bond. Polythene is the simplest example – it is formed from ethene:

Imagine a number of ethene molecules lined up:

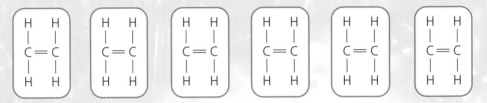

The double bonds open and form bonds with neighbouring molecules:

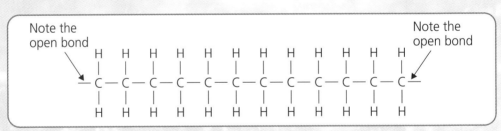

Note the open bond

Note the open bond

Usually, a question will just ask you to show 3 of the monomer units combined. Always remember to show an open bond at both ends of the section of polymer.

PVC is made from the vinyl chloride monomer:

H Cl
| |
C = C
| |
H H

Summary continued ➢

Summary *continued*

This means that every second carbon atom in the polymer has one chlorine atom bonded to it.

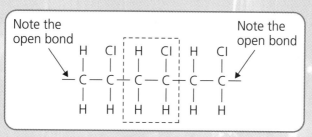

The bit in the dotted box is called the *repeating unit*, and you would draw it like this:

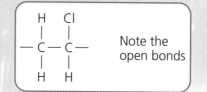

Polystyrene is made from styrene:

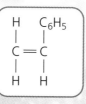

This means that every second carbon atom in the polymer has a C_6H_5– group bonded to it.

Addition polymers form by the linkage of carbon to carbon double bonds, and this means that the backbone of any addition polymer is made up of carbon atoms only.

Condensation polymerisation

Condensation polymers form from monomers with reactive groups of atoms (functional groups) at each end of the molecule. We've already met esters, so you should easily understand how *polyesters* form.

Imagine an alcohol with an – OH group at each end:

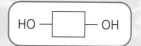

The square just stands for the rest of the molecule.

Imagine an acid with a – COOH group at each end:

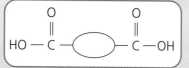

Summary *continued* ➤

Summary *continued*

The oval just stands for the rest of the molecule.

Imagine the acid and alcohol molecules lined up like this:

Condensation reactions, setting free H_2O, can take place indicated by the dotted ovals below:

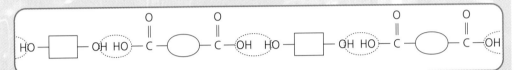

This gives a section of polymer as shown here:

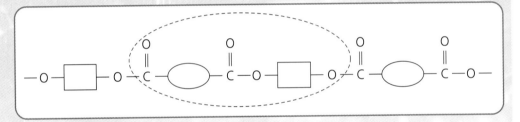

Its repeating unit is as shown by the large dotted oval. You should be able to spot ester links arranged along the backbone – this is a polyester. You can see that the backbone of this polymer is not composed of carbon atoms only.

Polyamides, like nylon and Kevlar, form from monomers like these:

and

The $-NH_2$ group is called the amino group. 2 hydrogen atoms and 1 oxygen atom are lost as water in a condensation reaction. The resulting polymer is very like the polyester above, but instead of ester links all along the polymer there are amide (or peptide) links which look like this:

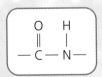

Summary *continued* ➤

Summary *continued*

Some plastics melt when heated – they are *thermoplastic*. Some are heated to make them set, and further heating does not melt them – they are *thermosetting*. Many plastics give off toxic fumes when they burn – most give off carbon monoxide (because they contain carbon). The chlorine in PVC results in HCl being produced. Polymers containing nitrogen can give off hydrogen cyanide, HCN, which is extremely toxic.

Learn a use for each common plastic – and remember that polyethenol is soluble in water, while Kevlar is very strong!

Example 14

The structure below shows a section of an addition polymer:

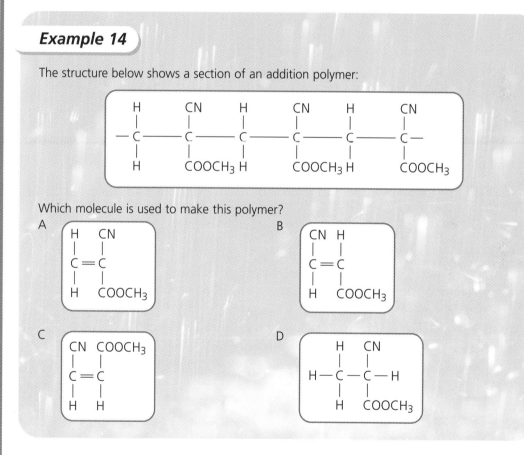

Which molecule is used to make this polymer?

Solution

You can tell that the polymer is an addition polymer because its backbone consists of carbon atoms only. It must therefore have a monomer with a carbon to carbon double bond. That rules out D. Note also that the – CN group and the – COOCH$_3$ group are on the same carbon atom in the polymer. They will therefore be on the same carbon atom in the monomers. This leaves **A** as the only possible monomer.

Example 15

Kevlar is a thermosetting polymer which can be used to make bulletproof vests.

The diagram below shows how the monomers used to make Kevlar link together:

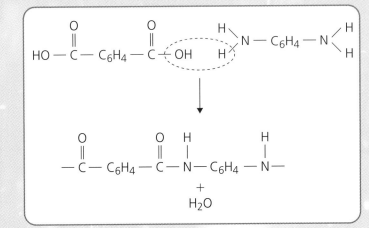

a)   What type of polymerisation takes place?

b)   Name the type of link formed.

c)   Why is it important that the molecules have functional groups at each end of the molecule?

d)   What property of Kevlar makes it suitable for use in bulletproof vests?

Solution

a)   Well, the backbone does not consist entirely of carbon atoms – a clue that this is a **condensation polymer**. Anyway, the appearance of water as one of the products should be a giveaway!

b)   It's an **amide link** – which you'll know, if you've learned your work.

c)   A simple alkanoic acid has just one – COOH group and a simple alkanol has just one – OH group. As a result, 1 alkanoic acid molecule can combine with 1 alkanol molecule to form a simple ester. Having a functional group at both ends of the molecule means that the acid and alkanol molecules **can combine endlessly to form long chains**.

   It's like people – having two hands they can all hold hands to form almost endless chains. If humans only had one hand they could only join up in pairs.

d)   All you need to know about Kevlar is that it's **strong**.

Example 16

Part of the structure of an addition polymer is shown below. It is made using two different monomers:

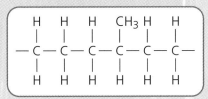

Which pair of alkenes could be used as monomers?

A   Ethene and propene

B   Ethene and butene

C   Propene and butene

D   Ethene and pentene

Solution

This isn't one of the easiest questions. Remember that alkenes always link up across the double bond. If you think of propene, but-1-ene and pent-1-ene like this:

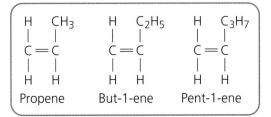

then you'll see that if they are involved in addition polymerisation then the polymer chain must have side chains of $-CH_3$ or $-C_2H_5$ or $-C_3H_7$. The only side chain in our polymer is $-CH_3$. All the other atoms attached to the backbone are hydrogen atoms. This suggests that the sample of polymer shown consists of an ethene linked to a propene linked to another ethene. The answer is **A**.

Natural products

Summary

There are many natural products but the Intermediate 2 course looks at products which are related to foods. These are *carbohydrates*, *proteins* and *fats and oils*.

Summary continued ➤

Summary *continued*

Carbohydrates

Carbohydrates are made of carbon, hydrogen and oxygen atoms. Hydrogen and oxygen are in the ratio of 2 atoms to 1 atom, respectively. The most common carbohydrates include *glucose* and *starch* made in plants by *photosynthesis*:

carbon dioxide + water → glucose + oxygen

Plants contain *chlorophyll*, which enables the plant to use the Sun's energy to carry out the reaction. Glucose molecules then combine to form starch.

Our bodies get energy from carbohydrates by the process of respiration, which is the reverse of photosynthesis.

You should have met glucose and fructose. They both have molecular formula $C_6H_{12}O_6$. You'll also have met sucrose and maltose, with molecular formula $C_{12}H_{22}O_{11}$. These are all sugars and all of them, except sucrose, give a positive reaction to Benedict's reagent.

Another important carbohydrate is starch, which is a condensation polymer of glucose units. Starch gives a negative result for Benedict's test but turns iodine a blue/black colour. Enzymes in our bodies break starch down to glucose in a hydrolysis reaction. You should remember that enzymes work best at particular pH values and at a temperature of around 37 °C. Starch can also be broken down to glucose in the lab by heating it with an acid.

Proteins

Proteins are what most of human bodies are made of – all your muscles are made of protein. Enzymes and many hormones are proteins too.

Proteins are condensation polymers. The monomers are *amino acids* with the general structure below:

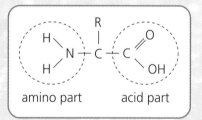

'R' stands for the rest of the molecule, which might be just a hydrogen atom or something more complicated. When amino acids link up by condensation they form the amide (or peptide) link which you've already seen in the section on polymers.

Summary *continued* ➤

Summary *continued*

Fats and oils

We need fats and oils in our diets, as a concentrated source of energy. We get them from animals, fish (marine oils) and plants.

Fats and oils are esters. They're built from an alcohol called *glycerol*. It has this structure:

Figure 3.5 Kippers have lots of marine oils

Make sure you can draw and recognise this molecule. The acids involved are alkanoic acids, often with about 16 carbon atoms. They're commonly known as *fatty acids*. Glycerol has 3 – OH groups so it can form 3 ester links to 3 molecules of acid. If the acid is unsaturated – that is, it has a carbon to carbon double bond – then an oil forms. If the acid is saturated then a fat forms.

We can turn oils into fats by adding hydrogen to the double bond. The process is called *hardening* in industry.

Example 17

Which one of the following is an example of a polymer?

A   Plant sugar

B   Animal fat

C   Marine oil

D   Vegetable protein

Solution

Think about each choice in turn. You're expected to know that typical sugar formulae are $C_6H_{12}O_6$ or $C_{12}H_{22}O_{11}$. That's certainly not big enough to be a polymer. Fats and oils are built from 1 molecule of glycerol condensed with 3 carboxylic acid molecules – again, hardly a polymer. So that leaves **D**, and you should remember that proteins are polymers of amino acids.

Example 18

What is the ratio of glycerol molecules to fatty acid molecules on the hydrolysis of a fat?

A   1 : 1
B   1 : 2
C   1 : 3
D   1 : 4

Solution

Just remember the structure of glycerol, shown on the previous page, and you'll remember that it can form 3 ester links. The ratio is 1 : 3 so the answer is **C**.

Example 19

Which one of the following correctly shows all the elements present in carbohydrates and proteins?

	Carbohydrates	Proteins
A	C, H and O	C, H and N
B	C and H	C, H and O
C	C, H, O and N	C and H
D	C, H and O	C, H, O and N

Solution

If you think about the word 'carbohydrate' it practically tells you what elements are present. The 'carbo' bit obviously means carbon! The 'hydrate' suggests water (as in dehydration) and water contains hydrogen and oxygen. The answer must be either A or D. Proteins are built of amino acids which contain carbon, oxygen, hydrogen and nitrogen (remember the word 'CHON') so the answer must be **D**.

If you look carefully at the question, you'll see that knowing CHON on its own would let you answer it correctly! If you think about it, these are also the atoms which make up the peptide link, so maybe you should remember them as CONH instead.

Example 20

The table shows the saturated and unsaturated fatty acid content of a fat and an oil.

Source	Fat / oil	% Fatty acids in substance	
		Saturated	Unsaturated
Animal	Chicken fat	68	32
Marine	Cod liver oil	25	75

Example continued ➤

Example 20 *continued*

a) What do fats and oils provide in our diet?

b) Name another source of fats and oils.

c) Why do oils have a lower melting point than fats?

d) How can oils be converted into hardened fats?

Solution

a) You should know that they are a **concentrated source of energy**.

b) As said earlier, there are three sources of fats and oils – animal, marine and vegetable (such as olive oil or sunflower oil). **Vegetables** is the answer.

c) This is to do with the fact that the acids which are involved in oils are unsaturated – that is, **they have carbon to carbon double bonds**.

d) This is done by getting rid of some of the double bonds in the oil. **Hydrogen is added** across the double bond in a process called 'hardening'.

Example 21

Which type of compound is represented by the structure shown?

A Amine

B Protein

C Amino acid

D Carboxylic acid

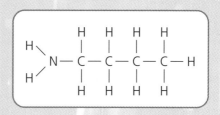

Solution

Well, if it was an acid then it would have a carboxyl group, – COOH. So it can't be C or D. Proteins are polymers of amino acids and the molecule shown certainly isn't a polymer. It's an amine. The – NH$_2$ group is the amino group, found in amines. **A** is the answer.

Example 22

Jelly is made from a protein called gelatine.

a) Name the kind of monomers which join together to form proteins.

Figure 3.6 Jelly beans – a source of protein (and sugar!)

Example continued ➤

Example 22 *continued*

A section of the gelatine structure is shown below:

$$-N-CH_2-C-N-CH_2-C-N-CH_2-C-$$

b) Circle a peptide link in this section of the gelatine structure.

c) Draw the structure of the monomer which makes this section of the gelatine molecule.

d) Papain, an enzyme found in pineapples, can hydrolyse gelatine. When the enzyme is heated to 100 °C it no longer hydrolyses the gelatine. Suggest why.

Solution

a) If you don't know by now that the monomers which make up a protein are **amino acids** you've got some work to do!

b) This should also be well known by now. The peptide link is:

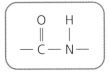

Remember, it contains all the elements found in every protein. So you know what to circle.

c) The monomers are amino acids, so that's what you have to draw. You'll remember that an amino acid has an amino group, $-NH_2$, at one end and a carboxyl group, $-COOH$, at the other. If you look closely at the gelatine structure above, you should spot that there are 2 peptide links. That means there are 3 amino acid units present, and they're all the same (though they needn't be). The amino acid you want is:

$$H-N-CH_2-C-OH$$

d) When enzymes are heated much above 37 °C they **undergo denaturation** – their structures are disrupted and they no longer work as enzymes.

Example 23

Which one of the following is the molecular formula for a carbohydrate?

A $C_6H_{14}O$ B $C_6H_{12}O_2$

C $C_6H_{12}O_4$ D $C_6H_{12}O_6$

Solution

There are two ways to get the answer to this. You need to know that glucose and fructose both have molecular formula $C_6H_{12}O_6$. In fact, they're isomers of each other. You also have to know, of course, that glucose and fructose are carbohydrates! So **D** is the answer.

You should also know that carbohydrates are made of carbon, hydrogen and oxygen atoms. The hydrogen and oxygen are in the ratio of 2 atoms to 1 atom, respectively. D is the only formula that has the hydrogen and oxygen in the correct ratio.

Example 24

Lipase is an enzyme which can catalyse the hydrolysis of fats in milk. Complete the diagram to show how the indicator colour would change after lipase was added to the test tube.

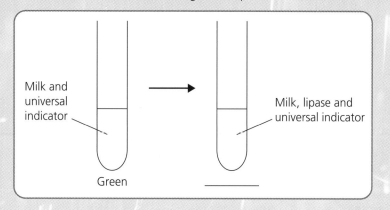

Milk and universal indicator

Green

Milk, lipase and universal indicator

Solution

This is really problem solving, but it's a question with lots of fat chemistry in it. First of all, you've got to recall that fats are esters of the alcohol glycerol – and you should be able to draw its structure – with long chain fatty acids. You also need to remember that hydrolysis breaks esters down into the acid and alkanol they were formed from. The effect of the lipase is to produce glycerol and acids from the fat. So there's acid present at the end of the experiment, which wasn't there before. All you need to know now is the colour that universal indicator turns in acid – and that's red. If you write '**red**' on the line, you'll get a mark!

Example 25

Carbohydrates are an essential part of our diet.

a)   Why are carbohydrates an important part of our diet?

b)   Name the elements present in carbohydrates.

c)   A student tested the carbohydrates glucose, sucrose and starch using iodine (Test 1) and Benedict's solution (Test 2).

Example continued ➢

Example 25 *continued*

Complete the table by identifying each carbohydrate.

Carbohydrate	Results	
	Test 1	Test 2
	Brown → blue/black	No change
	No change	No change
	No change	Blue → orange/red

Solution

a) Simple – you get **energy** from carbohydrates!

b) This was answered earlier. The name 'carbohydrate' suggests **carbon** and the elements of water – **hydrogen and oxygen**. Remember, proteins have CHON. Carbohydrates have the same elements as proteins without the nitrogen.

c) There's only one way to answer this and that's to know your stuff. The only food that iodine reacts with is starch, which it turns blue/black. Glucose gives a positive result for Benedict's test and sucrose doesn't. So your table should end up like this.

Carbohydrate	Results	
	Test 1	Test 2
Starch	Brown → blue/black	No change
Sucrose	No change	No change
Glucose	No change	Blue → orange/red

The last two examples have involved how test reagents like universal indicator and iodine and Benedict's reagent behave. It's useful to have a list of tests and their results – they're handy in both knowledge and understanding and in problem solving.

On the following page there is a list of tests – you should learn it. You could even make up a set of flash cards and test yourself.

Test	Distinguishes
Bromine solution	Alkenes will react with bromine decolourising it; alkanes do not react readily
Benedict's reagent	Positive result for reducing sugars – glucose, fructose and maltose; negative result for starch and sucrose
Starch	Iodine (gives blue/black colour)
Iodine	Starch (gives blue/black colour)
Addition of acid releases carbon dioxide	Carbonate
Orange/yellow flame test	Sodium compound
Lilac flame test	Potassium compound
Gas which burns with a 'pop'	Hydrogen
Gas which turns lime water cloudy	Carbon dioxide
Gas which relights a glowing taper	Oxygen
Ferroxyl indicator	Fe^{2+} ions (gives blue colour)
Ferroxyl indicator	OH^- ions (gives pink colour)

ACIDS, BASES AND METALS

Most of the new chemical reactions you'll meet in Intermediate 2 Chemistry are in this unit.

The unit covers the following topics.

◆ Acids and bases
◆ Salt preparation
◆ Metals

Acid and bases

The pH scale

Summary

The *pH scale* runs from below 0 to above 14. pH values below 7 are *acidic* and those above 7 are *alkaline*. pH 7 is *neutral*. The further away from pH 7 then the more acidic or alkaline a solution is.

Many oxides of non-metals dissolve in water to give acidic solutions, while oxides of metals dissolve in water to give alkaline solutions.

Watch out, though – lots of metal oxides don't dissolve in water (you can check this out using the Data Booklet) – and water, hydrogen oxide, is neutral!

Pure water, which is neutral with a pH of 7, contains equal numbers of H^+ and OH^- ions. These ions are formed in the breakdown of water molecules:

$$H_2O \rightleftharpoons H^+ + OH^-$$

This means that water molecules can form ions, and that the ions can recombine to form water molecules. When the breakdown and recombination occur at the same rate, we say that the system has reached *equilibrium*.

An acid solution contains a greater concentration of H^+ ions than of OH^- ions. An alkaline solution contains a greater concentration of OH^- ions than of H^+ ions. A neutral solution has the same concentration of both kinds of ion.

As an acid is diluted its pH rises toward 7 because the concentration of acid is getting less and less. In a similar way, as we dilute an alkali its pH falls toward 7.

You'll be expected to know about common acids such as vinegar, lemonade and Coke and common alkalis such as baking soda, detergents and bleach.

Example 1

Identify the oxide which will dissolve in water to give an alkaline solution.

A H_2O

B CO

C Na_2O

D CuO

Solution

If the oxide is to give an alkaline solution then it must be the oxide of a metal. Hang on – there are two metal oxides here, Na_2O and CuO. Remember that very few metal oxides dissolve in water at all. The Data Booklet tells you that Na_2O does dissolve in water and that CuO doesn't. So **C** is the answer.

It's worthwhile remembering that Group I metals, like Na, are called the alkali metals – this gives a clue about their behaviour. It's also worth remembering that apart from the metals in Groups I and II most metal oxides are insoluble in water. As a matter of interest, neither of the two non-metal oxides in the list is acidic. H_2O is water and CO doesn't dissolve in water so it hasn't got a pH.

Example 2

Which line in the table correctly describes what happens to a dilute solution of hydrochloric acid when water is added to it?

	pH	$H^+(aq)$ concentration
A	Increases	Increases
B	Increases	Decreases
C	Decreases	Increases
D	Decreases	Decreases

Solution

Acids have a pH of less than 7. Water has a pH of 7. So it's logical that adding water to acid moves the pH upwards toward 7 – or the pH increases. This means the answer is either A or B.

You also have to remember that acids contain lots of H^+ ions, and if you add water then they will become more dilute – their concentration decreases. This means that **B** is the answer.

Example 3

Which one of the following statements describes the concentrations of $H^+(aq)$ and $OH^-(aq)$ ions in pure water?

A   The concentrations of $H^+(aq)$ and $OH^-(aq)$ ions are equal.

B   The concentrations of $H^+(aq)$ and $OH^-(aq)$ ions are zero.

C   The concentration of $H^+(aq)$ ions is greater than the concentration of $OH^-(aq)$ ions.

D   The concentration of $OH^-(aq)$ ions is greater than the concentration of $H^+(aq)$ ions.

Solution

The easiest way to tackle this is to know the answer – that in pure water, the concentrations of the ions are equal. The answer is **A**. No one's ever told you that the concentration of either of these ions in water is zero. In C, the solution is an acid as there are more $H^+(aq)$ ions than $OH^-(aq)$ ions, while in D the solution is alkaline as there are more $OH^-(aq)$ ions than $H^+(aq)$ ions.

Concentration

Summary

Concentrations of solutions are measured in terms of *moles per litre*, or $mol\ l^{-1}$. It's important to understand that when we talk about 'moles per litre', we mean per litre of solution – not per litre of water used to make the solution. If we had 1 mole of sodium chloride and dissolved it in 1 litre of water then the volume would be more than 1 litre, because the sodium chloride would have some volume too.

$$\text{If concentration} = \frac{\text{number of moles}}{\text{number of litres}} \text{ then:}$$

◆   number of moles = concentration × number of litres;

$$\text{number of litres} = \frac{\text{number of moles}}{\text{concentration}} .$$

Example 4

0.5 mol of citric acid was dissolved in water and the volume made up to 250 cm³ by adding water. What was the concentration of the solution formed?

A   0.25 mol l^{-1}

B   0.5 mol l^{-1}

C   1.0 mol l^{-1}

D   2.0 mol l^{-1}

Solution

Because the concentration is in moles per litre, this is simply a case of scaling up 0.5 moles in 250 cm^3 (which is 0.25 litre) to the number of moles in 1.0 litre. This is 4 times the volume of 0.25 litre, so there will be 4 times as many moles present, that is, 4×0.5, which is 2 moles. The answer is **D**.

Example 5

Which of the following solutions contains the most dissolved solute?

A   100 cm^3 of a 4 mol l^{-1} solution

B   200 cm^3 of a 3 mol l^{-1} solution

C   300 cm^3 of a 1 mol l^{-1} solution

D   400 cm^3 of a 0.5 mol l^{-1} solution

Solution

Remember, number of moles = concentration $\times$ number of litres. You just have to multiply the volume in litres by the concentration to find the number of moles present in each solution. Remember that there are 1000 cm^3 in a litre. So:

A $0.1 \times 4 = 0.4$ mole

B $0.2 \times 3 = 0.6$ mole

C $0.3 \times 1 = 0.3$ mole

D $0.4 \times 0.5 = 0.2$ mole

B is the winner with 0.6 mole solute.

Strong and weak acids and bases

Summary

A solution of hydrochloric acid, HCl, doesn't contain very many HCl molecules. Most of the acid has split (*ionised* or *dissociated*) into H$^+$ ions and Cl$^-$ ions. On the other hand, ethanoic acid, CH$_3$COOH, is present mainly as CH$_3$COOH molecules and there are very few H$^+$ ions and CH$_3$COO$^-$ ions present. We say that HCl is a *strong acid* and that CH$_3$COOH is a *weak acid*. A strong acid gives lots of H$^+$ ions. A weak acid gives very few. In a similar way, a *strong base* gives lots of OH$^-$ ions and a *weak base* gives very few OH$^-$ ions.

You're expected to know that hydrochloric, nitric and sulphuric acids are strong acids. Any other acid you meet is likely to be a weak acid. Strong bases include sodium hydroxide and potassium hydroxide. The only weak base you're likely to meet is ammonium hydroxide. Generally, we think of bases as compounds which can *neutralise* acids. This includes metal hydroxides, oxides and carbonates.

Summary continued ➤

Summary *continued*

If we compare a weak acid and a strong acid (with the same concentrations) we find that the strong acid has a lower pH, a higher electrical conductivity and reacts faster than the weak acid. This is because there is a higher concentration of H^+ ions in the strong acid.

If we compare a weak base and a strong base (with the same concentrations) we find that the strong base has a higher pH, a higher electrical conductivity and reacts faster than the weak base. This is because there is a higher concentration of OH^- ions in the strong base.

Figure 4.1 The atmosphere of Venus contains the strong acids hydrochloric acid and sulphuric acid

Example 6

Which one of the following compounds is a base?

A   Potassium carbonate

B   Potassium chloride

C   Potassium nitrate

D   Potassium sulphate

Solution

You saw above that the three types of compound classified as bases are metal oxides, hydroxides and carbonates because they can all neutralise acids. So **A** is the answer.

Example 7

a)   In a solution of ethanoic acid only a few of the molecules break up to form ions. What does this indicate about ethanoic acid?

b)   In a solution of hydrochloric acid all of the hydrogen chloride molecules have broken up to form ions. Two properties of solutions of ethanoic acid and hydrochloric acid were compared. The results for ethanoic acid are shown in the table. Circle the appropriate words in the right-hand column of the table to show how the results for hydrochloric acid would compare with those for ethanoic acid.

Example continued ➤

Example 7 continued

	0.1 mol l^{-1} ethanoic acid	0.1 mol l^{-1} hydrochloric acid	
pH	4	Lower	Higher
Rate of reaction with magnesium	Slow	Lower	Higher

Solution

a) To answer this, you simply need to know what's meant by the terms 'weak' and 'strong'. **Ethanoic acid is a weak acid** because not all the molecules have broken down into ions.

b) Again, it's just a matter of learning your stuff. You're told that all the molecules have broken down into ions. Because there are more H$^+$ ions than in ethanoic acid, the pH will be **lower** and the rate of reaction **higher**. Easy marks, if you know your stuff.

Example 8

a) Citric acid is a weak acid. What is meant by a weak acid?

b) Hydrochloric acid is a strong acid. A student was given a bottle of 1 mol l^{-1} hydrochloric acid and a bottle of 1 mol l^{-1} citric acid. Describe an experiment that could be carried out to show that citric acid is a weak acid and hydrochloric acid is a strong acid. State the result that would be expected.

Solution

(a) Once again, make sure you can explain what's meant by the terms 'weak' and 'strong' when describing acids. In weak acids **only some of the molecules break down into ions**. In strong acids they all do.

(b) There are loads of experiments you can suggest. A strong acid contains more H$^+$ ions than a weak acid of the same concentration. As a result, it has a lower pH than the weak acid. It also has a higher electrical conductivity because it is the ions which carry the current. Because it is the H$^+$ ions which are responsible for the way an acid reacts, a strong acid will react faster than a weak acid with metals and compounds such as carbonates. Remember, you've got to be able to prove which acid is which, so there has to be a definite difference in behaviour when tested.

If you decide to measure pH then make sure you say that the **hydrochloric acid will have the lower pH** and say it's because there are more H$^+$ ions. If you decide to measure electrical conductivity then say that the **hydrochloric acid has a higher conductivity**, and give the reason (same reason!). If you decide to add a metal then make sure it's a metal (like magnesium) that will actually react with an acid – don't pick copper (why not?). Say what you'll see happening – magnesium will **release bubbles**

of hydrogen faster in hydrochloric acid than in citric acid. If you decide to add a metal carbonate to each acid, it's similar but this time the gas is carbon dioxide.

What you can't add to the acids are alkalis, such as sodium hydroxide, because no gas is given off (water and a salt are the products) so you won't see any difference in the way they react.

Salt preparation

Summary

When acids react with bases, the products include *salts*. As we said before, bases are compounds such as metal oxides, metal hydroxides and metal carbonates. Bases which dissolve in water are alkalis.

Suppose the acid is hydrochloric acid. The reactions of these compounds follow the following pattern:

metal oxide + hydrochloric acid ➔ metal chloride + water

metal hydroxide + hydrochloric acid ➔ metal chloride + water

metal carbonate + hydrochloric acid ➔ metal chloride + water + carbon dioxide

You can also make salts by reacting certain metals with acids. This reaction releases the gas hydrogen, which you can detect because it burns with a squeaky 'pop'.

Metal chlorides are examples of salts. A salt usually contains a metal ion and an ion from an acid. Salts made from hydrochloric acid contain *chloride ions*, those made from sulphuric acid contain *sulphate ions* and those made from nitric acid contain *nitrate ions*.

As well as metal ions, some salts contain ammonium ions, NH_4^+. Ammonium salts, especially ammonium nitrate, are useful as fertilisers because they contain the element nitrogen, which plants need in order to grow.

How to make salts

A salt can be made by adding an alkali to an acid. You add the alkali to the acid (or vice versa) until the pH is 7. You can tell when the pH is 7 by using an *indicator*.

If you're using an insoluble oxide or carbonate then you keep adding it to the acid until the reaction stops and there's some unreacted solid base left. You filter off the surplus and you're left with a solution of the salt. If you leave it to evaporate, crystals of the salt will eventually form.

If the salt is insoluble (you can check it in the Data Booklet) you can mix solutions of two soluble salts. Each salt contains one of the ions of the insoluble salt. If you want to make barium sulphate, which is insoluble, you can mix solutions of sodium sulphate and barium chloride. A *precipitate* of barium sulphate forms and it can be filtered off.

Example 9

Which one of the following compounds is a salt?

A   Potassium carbonate

B   Potassium chloride

C   Potassium oxide

D   Potassium hydroxide

Solution

Metal oxides, hydroxides and carbonates are bases. So the answer is not A, C or D. The answer is **B**, potassium chloride, as it contains a metal ion (potassium) and an ion from an acid (hydrochloric acid).

Example 10

Which one of the following solutions, when added to copper(II) chloride solution, will produce a precipitate? (You can use page 5 of the Data Booklet.)

A   Sodium hydroxide

B   Calcium bromide

C   Magnesium nitrate

D   Lithium sulphate

Solution

The first problem here is to decide whether the precipitate is a copper(II) compound or a chloride. Have a look at page 5 of the Data Booklet – down the chloride column. You'll see that the only insoluble chloride shown is silver chloride, so the precipitate isn't a chloride. It must be a copper(II) compound.

Look along the copper (II) row – copper bromide, copper nitrate and copper sulphate are all 'vs' – very soluble. Copper hydroxide is 'i' – insoluble. So **A** is the answer. When sodium hydroxide solution is added to a solution of copper(II) chloride you get a precipitate of copper(II) hydroxide.

Ionic equations and spectator ions

Summary

We often show neutralisation reactions and precipitation reactions by means of *ionic equations*. An example will help to make things clear. Think about the reaction between hydrochloric acid and sodium hydroxide. In words:

Summary continued ➤

Summary *continued*

hydrochloric acid + sodium hydroxide → water + sodium chloride

As a formula equation, it's:

$$HCl + NaOH \rightarrow H_2O + NaCl.$$

As an ionic equation it's:

$$H^+(aq) + Cl^-(aq) + Na^+(aq) + OH^-(aq) \rightarrow H_2O + Cl^-(aq) + Na^+(aq)$$

This shows the ions which are actually doing something – in this case, it's the H^+ and the OH^- ions.

The ions which aren't really doing anything are called *spectator ions*. You'll often find questions that ask you to rewrite an equation without the spectator ions, or ask you to show which ions are spectator ions.

Example 11

These two questions refer to the following reaction – when lead(II) nitrate solution is added to sodium iodide solution, a precipitate of lead(II) iodide is formed.

a) A sample of the precipitate can be separated from the mixture by

   A   condensation

   B   distillation

   C   evaporation

   D   filtration.

b) The equation for the reaction is:

$$Pb^{2+}(aq) + 2NO_3^-(aq) + 2Na^+(aq) + 2I^-(aq) \rightarrow Pb^{2+}(I^-)_2(s) + 2Na^+(aq) + 2NO_3^-(aq)$$

The spectator ions present in this reaction are

A $Na^+(aq)$ and $NO_3^-(aq)$

B $Na^+(aq)$ and $I^-(aq)$

C $Pb^{2+}(aq)$ and $NO_3^-(aq)$

D $Pb^{2+}(aq)$ and $I^-(aq)$.

Solution

a) In the first part of the question, a precipitate is separated from a liquid mixture by filtration, **D**. Just learn it.

b) In the second part, first of all notice that the $Pb^{2+}(aq)$ and $I^-(aq)$ ions have ended up as $Pb^{2+}(I^-)_2(s)$. The (s) tells you that the lead iodide has formed as a solid – a precipitate. The $2Na^+(aq)$ ions and the $2NO_3^-(aq)$ ions were there at the start and they're there at the end – they are unchanged. These are the spectator ions, so **A** is the answer.

Sometimes the question is asked in the following way.

Example 12

A student added a solution of lead(II) nitrate to a solution of potassium iodide to form a precipitate of lead(II) iodide. The equation for the reaction is:

$$Pb^{2+}(aq) + 2NO_3^-(aq) + 2K^+(aq) + 2I^-(aq) \rightarrow Pb^{2+}(I^-)_2(s) + 2K^+(aq) + 2NO_3^-(aq)$$

a)   Rewrite the equation showing only the ions which react.

b)   What term is used to describe the ions which do not react?

Solution

a)   The ions which react are those that look different on the right of the arrow when compared with their appearance on the left of the arrow. The $Pb^{2+}(I^-)_2(s)$ on the right is obviously different from the $Pb^{2+}(aq)$ and $2I^-(aq)$ on the left, so the equation should be written

$$Pb^{2+}(aq) + 2I^-(aq) \rightarrow Pb^{2+}(I^-)_2(s)$$

b)   The ions which don't react are called **spectator ions** – in this case, the $2K^+(aq)$ and the $2NO_3^-(aq)$.

Sometimes, you'll see a question like this, in which you're asked to circle the spectator ions. In this case, your answer would look like this:

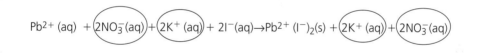

Make sure your circles take in all of the formula, just to be safe!

Volumetric titrations

Summary

If we don't know the concentration of an acid, we can measure out a certain volume of it and neutralise it using an alkali of known concentration. An indicator tells us when the acid has been neutralised.

From the volume of alkali required we can then calculate the concentration of the acid. If we have an acid of known concentration then we can use it to find the concentration of an alkali in the same way.

Example 13

A solution of sodium hydroxide can be used to find the concentration of a solution of sulphuric acid. The equation for the reaction is:

$$H_2SO_4 + 2NaOH \rightarrow Na_2SO_4 + 2H_2O$$

In one experiment, 20 cm^3 of a H_2SO_4 solution required 10 cm^3 of 0.1 mol l^{-1} NaOH for neutralisation.

Calculate the concentration of the H_2SO_4 solution. Show your working clearly.

Solution
There's more than one way of doing this kind of calculation.

Method 1
Work out the number of moles of NaOH used.

Moles = volume (litres) × concentration (mol l^{-1}) = 0.01 × 0.1 = 0.001

The equation tells us that 1 mole of H_2SO_4 reacts with 2 moles of NaOH.

Therefore, there were 0.0005 mole of H_2SO_4 present.

This was contained in 20 cm^3, which is 0.02 litre.

If there is 0.0005 mole in 0.02 litre, then the concentration of the H_2SO_4 is

$$\frac{0.0005}{0.02} = 0.025$$

The answer is **0.025 mol l^{-1}**.

Method 2
We can call this method the 'mole ratio' method. The mole ratio between acid and alkali is 1 : 2

$$H_2SO_4 + 2NaOH \rightarrow Na_2SO_4 + 2H_2O$$

1 mole   2 moles

The number of moles is found by multiplying the volume, V (litres) by the concentration, M (moles per litre). We'll use the subscript 'a' to stand for acid and the subscript 'b' to stand for alkali (base). Now we can write:

$$H_2SO_4 + 2NaOH \rightarrow Na_2SO_4 + 2H_2O$$

1 mole   2 moles

$$V_aM_a \qquad V_bM_b$$

Then we can say that $\dfrac{V_aM_a}{V_bM_b} = \dfrac{1}{2}$

We can then cross-multiply and get $2V_aM_a = V_bM_b$

We can rearrange this to give $M_a = \dfrac{V_b M_b}{2V_a} = \dfrac{0.01 \times 0.1}{2 \times 0.02} =$ **0.025 mol l^{-1}**

Method 3

We can call this method the 'PVC' method! As before, V is the volume of acid and alkali, and C is the concentration in mol l^{-1}. We use 'C' in this method rather than 'M' because it's easier to remember – but you could learn it as the PVM method if you like!

P is the 'power' of the acid or alkali. In the case of an acid, it's the number of H$^+$ ions present. In the case of an alkali it's the number of OH$^-$ ions present. So P for H_2SO_4 is 2 and for NaOH, it's 1. We write:

$$P_a V_a C_a = P_b V_b C_b$$

Then $2 \times 0.02 \times C_a = 1 \times 0.01 \times 0.1$

So $0.04 C_a = 0.001$ giving $C_a = \dfrac{0.001}{0.04} =$ **0.025 mol l^{-1}**.

This method is useful because you don't need the balanced equation for the reaction. You just need the formula for the acid and the base.

> **!** **Some acids, like ethanoic acid (CH$_3$COOH), can cause confusion because although it has 4 hydrogen atoms, only one of them can give H$^+$ ions, so its power is 1 and not 4.**

⇨ *Metals*

An order of reactivity

Summary

If we add different metals to an acid we find that some metals react quickly, some react slowly and some don't react at all. We can arrange them in an *order of reactivity*.

It turns out that we get the same order of reactivity no matter what we are making the metal react with – acid, water or oxygen. The reactivity depends on how easily the metal can lose electrons – the first step in any reaction of a metal. In the Data Booklet, you'll find the common metals included in the *electrochemical series* (ECS) from the most reactive metals at the top (the ones that lose electrons easily), to the least reactive metals at the bottom (the ones that don't lose electrons easily). It's worth remembering that all the

Figure 4.2 Is this a wind-up?

Summary *continued* ➤

Summary *continued*

metals above hydrogen will react with acids, while the top five – lithium to magnesium – will all react with water.

If we connect two different metals together in a conducting solution then an electric current flows between them, because one of the metals will lose electrons more easily than the other. The electrons flow from the metal higher in the ECS to the metal lower in the ECS. The further apart the metals are in the ECS the bigger the *voltage* produced. We can set up *cells*, where we have different metals in solutions of their ions connected by an ion bridge to complete the circuit. Cells don't have to involve metals only. Non-metals can also gain and lose electrons so they can take part in cells too.

Displacement reactions happen when you add a more reactive metal to a solution containing ions of a less reactive metal. If you add magnesium to a solution of copper ions, the magnesium loses electrons and forms Mg^{2+} ions, while the Cu^{2+} ions gain electrons and turn into copper metal. We show this using *ion–electron equations*:

$$Mg \rightarrow Mg^{2+} + 2e^-$$

$$Cu^{2+} + 2e^- \rightarrow Cu$$

The loss of electrons is called *oxidation* and the gain of electrons is called *reduction*. You can decide which of the equations shown above is which! When both processes happen at the same time then it's called a *redox* reaction.

Example 14

Which one of the following reacts with dilute hydrochloric acid to give hydrogen gas?

A   Copper

B   Gold

C   Magnesium

D   Mercury

Solution

You don't have much choice here! Pick the most reactive metal – the one highest in the ECS. Magnesium! The answer is **C**. The reaction happens because acids contain H^+ ions. Because magnesium is above hydrogen in the ECS, the magnesium will lose electrons and the H^+ ions in the acid will gain them. The ion–electron equations are:

$$Mg \rightarrow Mg^{2+} + 2e^- \text{ (oxidation)}$$

$$2H^+ + 2e^- \rightarrow H_2 \text{ (reduction)}$$

Example 15

Which one of the following metals will react with zinc chloride solution?

A   Copper

B   Gold

C   Iron

D   Magnesium

Solution

Again, the only thing you can do is go for the most reactive metal – the answer is **D**. The key thing is that the metal needs to be above zinc in the ECS – that means it's magnesium again. The reaction happens because the solution contains Zn^{2+} ions. Because magnesium is above zinc in the ECS, the magnesium will lose electrons and the Zn^{2+} ions in the solution will gain them. The ion–electron equations are:

$$Mg \rightarrow Mg^{2+} + 2e^- \text{ (oxidation)}$$

$$Zn^{2+} + 2e^- \rightarrow Zn \text{ (reduction)}$$

The other metals – copper, gold and iron – hold on to their electrons too tightly for the zinc ions to take them.

Example 16

A student set up the following cell.

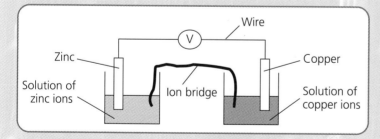

The reactions taking place at the electrodes are:

$$Cu^{2+} + 2e^- \rightarrow Cu$$

$$Zn \rightarrow Zn^{2+} + 2e^-$$

a)   On the diagram, clearly mark the path and the direction of electron flow.

b)   Combine the two ion–electron equations for the electrode reactions to produce a balanced redox equation.

c)   What is the purpose of the ion bridge?

Solution

a) Watch out! Electrons can only flow through metals (and graphite) so that means that the electrons must be flowing through the wires. They will travel **from the metal which loses them (the zinc) to the metal which gains them (the copper)**. What you draw must look like this:

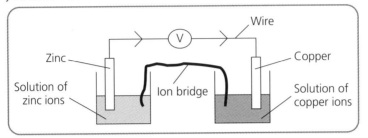

You must put your arrows **on** the wire – not above it and certainly not below it – because that might be taken to mean that the electrons are flowing through the ion bridge. Keep your arrows on the horizontal part of the wire – don't take them down into the beakers.

b) The ion–electron equations are:

$$Cu^{2+} + 2e^- \rightarrow Cu$$

$$Zn \rightarrow Zn^{2+} + 2e^-$$

To combine them, cancel out the electrons in equal numbers on both sides of the equations. The electrons are already present in equal numbers in this example. If they weren't, you'd have to multiply one (or both) of the equations to equalise them. You then add what's left and get:

$$Cu^{2+} + Zn \rightarrow Cu + Zn^{2+}$$

c) The usual acceptable answers to this question are '**to complete the circuit**' or '**to allow ions to flow**'.

Example 17

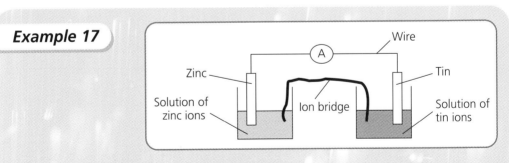

In the cell shown above, electrons flow through

A   the solution, from tin to zinc

B   the solution, from zinc to tin

C   the wire, from tin to zinc

D   the wire, from zinc to tin.

Solution

You have to remember what electrons do – they flow through metals. That means you must pick C or D. They flow from the metal higher in the ECS to the metal lower in the ECS – in this case from zinc to tin. The answer is **D**.

Extracting metals from their ores

Summary

Ores are naturally occurring compounds of metals that are found in the Earth. Many of these are metal oxides, which is not surprising when you think how much oxygen is present on Earth. Some ores are sulphides and others are silicates.

In order to extract metals, which are useful, from their ores – which are not useful in themselves – we need to remove the oxygen (or other elements with which the metal is combined). Just how easy this is depends on how reactive the metal is – that is, where the metal is in the ECS. The higher the metal is in the ECS then the more reactive it is and the more tightly it will try to hold on to the oxygen with which it's combined.

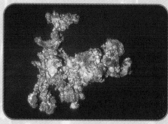

Figure 4.3 This silver nugget was found 'native' – that is, as pure metal

Near the foot of the ECS, the metals are so unreactive that they can be found as pure metals, not combined with anything else. If these very unreactive metals are found combined with oxygen then simply heating the oxide will split it into metal and oxygen.

Metals around the middle of the ECS can be made by heating their oxides with carbon or carbon monoxide. Iron is made in a *blast furnace* by heating iron oxide with coke (a form of carbon). Hot air is blown in at the foot of the furnace. This reacts with coke to form carbon dioxide. The carbon dioxide reacts with more coke to form carbon monoxide, which in turn removes the oxygen from the iron oxide. The equations for these processes are:

$$C + O_2 \rightarrow CO_2$$

$$CO_2 + C \rightarrow 2CO$$

$$Fe_2O_3 + 3CO \rightarrow 2Fe + 3CO_2$$

You're expected to know these equations!

Metals higher in the ECS hold on to their oxygen too tightly for this method to work. Then we have to use *electrolysis* of the molten metal oxide.

Metal	Method of extraction
Lithium, potassium, calcium, sodium, magnesium, aluminium	Electrolysis
Zinc, iron, nickel, tin, lead, copper	Heat with carbon or carbon monoxide
Silver, mercury, gold	Found pure, otherwise heat alone

Summary continued ➢

HOW TO PASS INTERMEDIATE 2 CHEMISTRY

Summary *continued*

Just remember the cut-offs between aluminium and zinc, and between copper and silver in the ECS!

Questions on extracting metals from their ores usually turn up in the multiple-choice part of the paper.

Example 18

Some metals can be obtained from their metal oxides by heat alone.

Which one of the following oxides would produce a metal when heated?

A   Calcium oxide

B   Copper oxide

C   Silver oxide

D   Zinc oxide

Solution

If you've remembered the cut-offs, you'll pick silver. Otherwise, go for the metal which is lowest in the ECS – again silver – so the answer is **C**.

Example 19

Which one of these metals must be obtained from its oxide by electrolysis?

A   Iron

B   Silver

C   Mercury

D   Aluminium

Solution

If you've remembered the cut-offs, you'll pick aluminium. Otherwise, go for the metal which is highest in the ECS – again aluminium – so the answer is **D**.

Corrosion

Summary

Corrosion takes place when the surface of a metal reacts with chemicals in its environment. This can change the appearance of the metal and make it less attractive. It can also weaken the metal so that it cannot do the job it is meant to do.

When iron corrodes, we call it *rusting*. Rusting is caused by the reaction of iron with oxygen and water. The reactions involved are:

$$Fe \rightarrow Fe^{2+} + 2e^-$$

$$H_2O + \tfrac{1}{2}O_2 + 2e^- \rightarrow 2OH^-$$

The Fe^{2+} ions then lose another electron to form Fe^{3+}. Fe^{3+} ions then react with OH^- ions to form $Fe(OH)_3$ which is a component of rust. *Ferroxyl indicator* is used to test for the presence of Fe^{2+} ions (it turns blue) and OH^- ions (it turns pink).

Rusting is accelerated by salt and acids – e.g. acid rain. Because rusting involves iron losing electrons, anything that makes it lose electrons faster speeds up rusting. This includes connecting iron to a metal lower in the ECS. Such a metal has a stronger attraction for electrons than iron and pulls electrons away from iron.

We can slow down rusting by giving electrons to the iron. We can do this by attaching a metal higher in the ECS such as magnesium (*sacrificial protection*) or by coating the metal with zinc (*galvanising*). Connecting the iron to a negative electrical terminal also slows rusting because the terminal supplies electrons (*cathodic protection*).

We can also slow rusting by covering the iron with paint, oil, plastic or another metal (e.g. zinc, which is used to galvanise iron). The covering simply keeps oxygen and water away from the surface of the iron. If we use a metal lower in the ECS to cover the iron then we must make sure that the coating isn't damaged in any way, otherwise oxygen and water get to the iron and it rusts faster than normal because the other metal removes electrons from the iron. If the covering is zinc then it still protects iron if it is damaged because it is above iron in the ECS.

Figure 4.4 Cathodic protection can now be solar powered!

Example 20

a) When iron rusts, iron(II) ions are formed. Write the ion–electron equation for the formation of iron(II) ions from iron atoms.

b) Name the solution used to test for iron(II) ions.

c) The apparatus and chemicals shown below can be used to show that both oxygen and water are required for rusting.

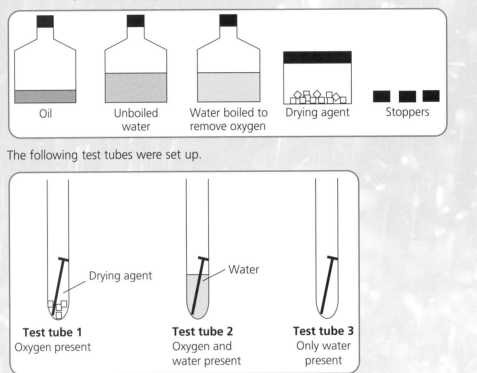

The following test tubes were set up.

Test tube 1
Oxygen present

Test tube 2
Oxygen and water present

Test tube 3
Only water present

Complete and label the diagram of test tube 3.

Solution

a) No problem – you've learned your equations and write:

$$Fe \rightarrow Fe^{2+} + 2e^-$$

b) No problem again – it's **ferroxyl indicator**. Make sure you know what colours it turns for Fe^{2+} (blue) and OH^- (pink).

c) This is probably an experiment you've done or had shown to you. When you boil water it gets rid of any oxygen dissolved in it. This isn't the same oxygen as the oxygen in the formula H_2O. It's the oxygen that fish use – on a hot summer day the heat may have driven the dissolved oxygen out of the water in a shallow pond and fish find it hard to survive.

If you add the water that has been boiled to tube 3, and leave the surface exposed to air, then air will dissolve in the water again so that the nail is exposed to both water and oxygen. To stop this happening, you add oil to the tube. It floats on the surface of the water and prevents air getting to the water. What you have to draw is this:

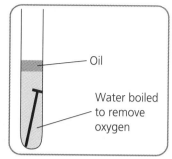

Oil

Water boiled to remove oxygen

Example 21

Oil rigs are made from steel. The oil rigs have zinc blocks attached to them to help to prevent rusting. The zinc is oxidised, protecting the steel.

a) Write the ion–electron equation for the oxidation of zinc.

b) Why would tin blocks not prevent rusting?

Solution

a) Oxidation is the loss of electrons, so the electrons have to appear on the right-hand side of the equation. You write:

$$Zn \rightarrow Zn^{2+} + 2e^-$$

It's very unlikely that you'll ever be asked to write an ion–electron equation that's not in the Data Booklet. In the Data Booklet, they're written as reduction equations – so in this case you'd have to find the equation for zinc and reverse it.

b) **Tin is below iron in the ECS**. It has a stronger attraction for electrons than iron has. Therefore it would attract electrons away from iron, speeding up rusting. The iron would be sacrificed to protect the tin.

Example 22

Iron is used to make ships. Sometimes, when a ship docks in a harbour, it is connected to the negative terminal of a power supply. A lump of scrap iron placed in the sea is connected to the positive terminal. This is used to prevent the iron in the ship from rusting.

a) How does connecting the ship to the negative terminal of the power supply prevent the iron in the ship from rusting?

b) Why is sea water able to complete the circuit?

Solution

a) When iron rusts it loses electrons. **The negative terminal of the power supply provides the iron with electrons**, slowing down the rusting. The electrons are taken from the lump of scrap iron which is connected to the positive electrode so the scrap iron will rust faster than normal.

b) **Sea water contains lots of ions** – for example Na^+ and Cl^-. The ions allow it to conduct electricity.

Example 23

The coatings on four strips of iron were scratched to expose the iron metal. The strips were placed in salt solution.

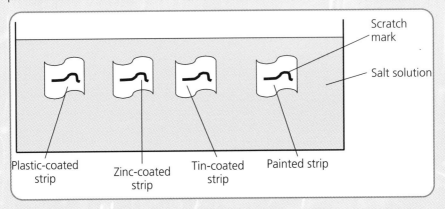

The iron strip which would have rusted most quickly was the one which was

A   plastic coated

B   zinc coated

C   tin coated

D   painted.

Solution

Paint and plastic don't have any effect on the iron, other than to keep out oxygen and water. If these coatings are scratched, the iron underneath will then rust at a normal rate. The salt solution will speed up rusting, but that's the same for all four strips.

The zinc-coated strip won't rust because although the zinc has been scratched, zinc is above iron in the ECS so it will let its electrons go more easily than iron. That means that iron is protected from rusting because it gets electrons from zinc.

The tin-coated strip will rust faster if the tin is scratched. Oxygen and water will reach the iron, which will begin to rust. Tin is lower than iron in the ECS and will attract electrons away from the iron. This makes it rust faster – so the answer is **C**.

PROBLEM SOLVING

Problem-solving questions sometimes draw on aspects of chemistry and chemical techniques you'll never have heard of. When you're faced with a question of this type, you have to read it carefully a couple of times to make yourself quite clear about what the question is telling you and asking you.

Because it might involve something you have never met, it will start off by setting the scene and this may take several lines. You must read it all carefully. Then try to be clear about what you are being asked to do. If you have read the introduction carefully then you should manage the questions without too much difficulty.

Many of these questions have nothing at all to do with the content of your chemistry course but are completely self-contained. They are included because they meet the requirements for the different kinds of problem solving which you're suppose to be able to do. However, many of the questions which are mainly Problem Solving have marks for Knowledge and Understanding as well, so it's unlikely that any long question will be directed at purely Problem Solving.

There are various kinds of problem-solving questions, and we'll have a look at some examples of each.

Graphs

If you're doing Intermediate 2 in fifth year, you might have done Standard Grade in third and fourth years. Your general level paper probably asked you to draw a bar chart. There aren't any bar charts in Intermediate 2 but you might be asked to draw a line graph. That's a lot harder. There are some useful tips which will help you.

Scales and axes

Be user friendly (to yourself as well as whoever marks the paper!) – pick a scale that goes up in fives or tens (or multiples of ten). Don't go up in fours or sixes (or sevens for that matter). It's really hard to plot the points accurately if you do that. Also, make sure that your scales will let you use at least half of the graph paper both vertically and horizontally. If you don't then you're likely to lose half a mark.

Make sure you label the axes and put the correct units in. You have to decide what should be plotted on the vertical axis and on the horizontal axis. You might have heard of 'independent variables' and 'dependent variables'. The *independent variable* is the factor you control. It should be on the horizontal axis. It's a bit like 'cause and effect' – the cause goes on the horizontal axis and the effect, the *dependent variable*, goes on the vertical axis. If you're plotting the rate of reaction against temperature, the temperature goes on the horizontal axis and the rate goes on the vertical axis. This is because the temperature affects the rate, not the other way round.

Example 1

Carbohydrates in plant waste can be digested by bacteria in the soil to produce biogas.

The table shows how the number of bacteria in soil is affected by the pH of the soil.

pH of soil	Number of bacteria/ millions per gram of soil
4.0	2.0
4.5	3.0
5.5	8.0
6.0	5.0
6.5	4.0

Draw a line graph of these results.

Solution

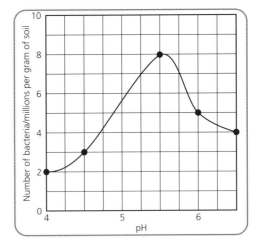

In this case, the independent variable (the factor the investigator can control) is the pH. The pH scale should be on the horizontal axis.

Watch out! The pH scale isn't in even jumps – most of the increases are 0.5 of a pH unit, but there's one jump of a whole pH unit. Be alert for this.

The origin

You have to decide if the line should pass through the origin – that's the 0, 0 point. Sometimes you are given data that include zero on the horizontal axis. If you're not then it's safer to avoid the origin. In this example, you've no real idea how many bacteria would be present at pH values down to zero, so your line would start at pH 4.0 and end at 6.5.

Join the points – or what?

You have to decide what kind of line to use in a line graph. If the points look as if they form a straight line then use a best-fit line which allows the points to be spread equally on either side of the line. If it looks like a smooth curve (like lots of graphs of reaction rate) then use a best-fit curve. If it's neither (like this example) you might just have to join the points!

Apparatus questions

Some problem-solving questions ask you to draw apparatus to carry out a simple experiment, or to complete a diagram of apparatus. Others give you diagrams of bits of apparatus and ask you how they should be put together. The important thing to realise is that you're **not** asked to invent any **new** apparatus. It will involve apparatus you should have seen many times in the lab. Let's look at a few apparatus questions.

Example 2

Lavender oil is extracted from the flowers by steam distillation. The flowers are put into a flask with a little water. Steam from a steam generator is blown through them to extract the oil. The mixture of lavender oil and steam distils over. It is condensed and collected. The pieces of apparatus used to carry out this steam distillation are shown here.

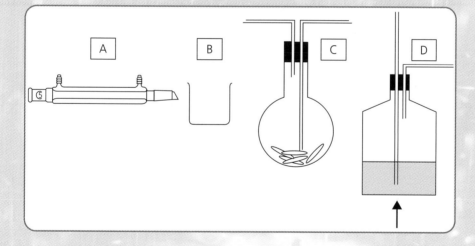

Put a letter in each box below to show the order in which the pieces of apparatus should be arranged to obtain the mixture.

Solution

You need to make sure that you read the words at the start of the question carefully. It mentions a 'steam generator'. To make steam you need water and heat. Is there anything that contains water and is being heated? D looks like the steam generator, so D will go in the first box. The steam is blown through the lavender flowers in a flask. That must be C, so C goes in the second box. Then you are told that the lavender oil is condensed and collected. You should recognise A as a condenser – cold water circulates round an inner glass tube to condense vapour in the tube. The beaker B must be for collection. So your answer is:

Example 3

Candle wax is a hydrocarbon. Blue cobalt chloride paper and lime water can be used to detect products formed when candle wax is burned.

Complete and label the apparatus below to show the arrangement you would use.

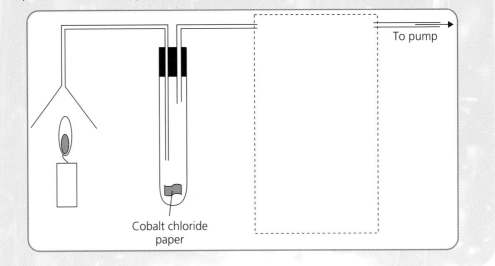

Cobalt chloride
paper

Solution

When you burn a hydrocarbon, carbon dioxide and water are produced. Cobalt chloride paper changes from blue to pink in the presence of water. The missing apparatus obviously must involve lime water with gases bubbling through it. You have to insert the following:

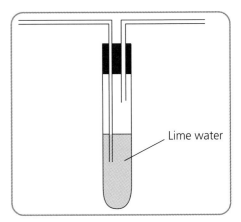

Lime water

It's really important that what you draw will actually work. In this example, the tubes in the test tube have to be the right length. Look at some of the ways you could go wrong!

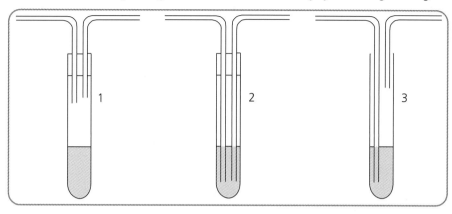

In the first diagram, the inlet tube (on the left) is too short. This means that the carbon dioxide doesn't have the chance of passing through the lime water, so it will not be detected.

In the second diagram, the outlet tube (on the right) is too long. This means that as gases enter the tube lime water is forced up the outlet tube and will end up in the pump, which might not do it much good! If you used a test tube with a side arm, this problem wouldn't arise and it avoids the problem very neatly.

In the third diagram, something important is missing. You've spotted it! There's no stopper in the test tube. The pump can't suck gases through the system unless it is properly sealed.

Example 4

In some countries, cow dung is fermented and the mixture of gases produced, known as biogas, is used as a fuel. The mixture contains a small amount of carbon dioxide.

The percentage of carbon dioxide in a biogas sample can be found by experiment. Part of the apparatus required is below. The gas mixture is forced through potassium hydroxide solution. This absorbs any carbon dioxide but not methane.

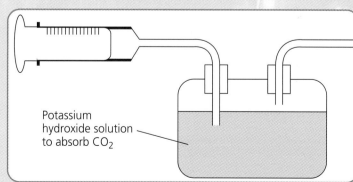

Potassium hydroxide solution to absorb CO_2

Complete the diagram to show all the apparatus which could be used to carry out this experiment.

Solution

All you have to do here is add another gas syringe on the right-hand side, like the one on the left, to collect the methane. But watch out – as in the example above, you must be sure not to close off any tube which should be open. Note also that you have to know how much methane you have collected (to work out the percentage of carbon dioxide). The gas syringe on the left-hand side has a scale on it. So, the one you draw on the right must also have a scale, otherwise you will not know how much gas you have collected – and you may not collect full marks for the question either!

Another way of measuring the volume of methane would be to collect the gas over water in an inverted measuring cylinder as shown in Figure 5.1. However, you must take care that your diagram doesn't include the common error of showing the delivery tube passing through the side of the water container or the measuring cylinder. It must pass over the rim of the water container and under the mouth of the measuring cylinder as shown in the diagrams.

Correct   Wrong!

Figure 5.1 Collecting gases by displacement of water

Spotting patterns and making predictions

You'll find questions of this type in almost every Intermediate 2 paper. It's a pattern!

Example 5

Gases can be liquefied by increasing the pressure – but above a certain temperature it is not possible to do this. This temperature is known as the critical temperature. The critical temperatures of some alkanes are shown below.

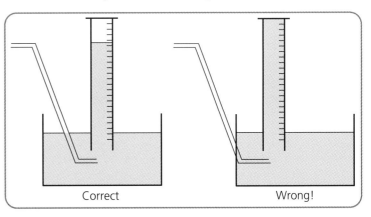

97 °C

152 °C

Example continued ➤

Example 5 *continued*

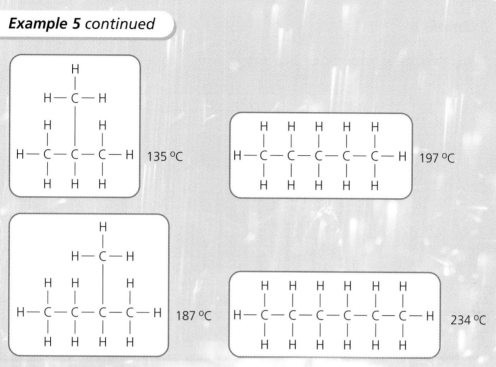

a) Describe the trend in critical temperatures for the straight chain alkanes.

b) Predict the critical temperature of the alkane:

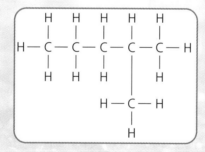

Solution

a) This is pretty straightforward. You should spot four straight chain alkanes in the table – from propane, with three carbons, to hexane, with six. You should see that their critical temperatures increase steadily, so your answer should be something like '**The more carbon atoms the higher the critical temperature**'.

b) If you look carefully at the table, you'll see that butane and pentane also have branched isomers in the table. The critical temperature of butane's branched isomer is lower than that of the straight chain isomer. In a similar way, the critical temperature of pentane's branched isomer is lower than that of straight chain pentane. The molecule in the question is a branched isomer of straight chain hexane, so its critical temperature must be lower than 234 °C. Its critical temperature will also be higher than that of pentane's branched isomer, 187 °C. So any answer **between 187 °C and 234 °C** will do.

Example 6

Energy is required to remove an electron from an atom. The graph shows the energy required to do this for the first 19 elements.

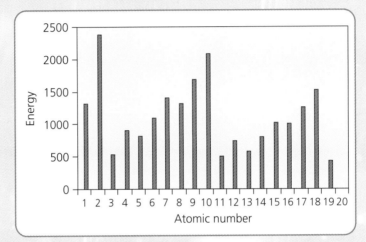

a) Describe what happens to the energy required going down a group in the Periodic Table.

b) Describe the general trend in the energy required going across the row from sodium to argon.

c) Draw a bar on the graph to show the energy you would expect to be required for the element with atomic number 20.

Solution

a) First you have to remember that a group is a vertical column of elements, such as Group 1 – lithium, sodium and potassium etc. or Group 0, the noble gases.

Suppose you choose Group 0. The first three elements are helium, neon and argon. Their atomic numbers are 2, 10 and 18 respectively. Look at the bars for these elements on the energy graph. They get shorter as you go down the group. So the answer is something like '**As you go down the group, less energy is required to remove an electron**'.

Check it out with another group, say Group 1, with atomic numbers 3, 11 and 19. The trend is the same but less obvious.

b) You're being asked for the general trend from element 11 to element 18. Be warned – when you're asked for a general trend, you're not being asked for a blow-by-blow account. Don't say 'It goes up to magnesium then falls at aluminium then goes up again to silicon and so on.' All you need to say is '**It goes up**'.

c) Element 20 is in Group 2 (it's calcium). The two other Group 2 elements on the graph are beryllium (4) and magnesium (12). Because the energy decreases as you go down a group, it must be less than magnesium's energy at 12. Because it rises as you go from Group 1 to Group 2, it must be higher than potassium's energy at 19. It could look like this.

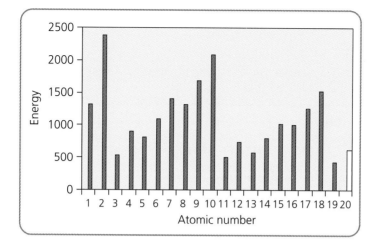

PROBLEM SOLVING

Example 7

Using a suitable catalyst and heat, alkanes can be made from alkanoic acids:

$$\text{alkanoic acid} \;\rightarrow\; \text{alkane} \;+\; \text{carbon dioxide}$$

Complete the table to show which alkanoic acid could be used to produce butane.

Alkanoic acid	Alkane
ethanoic acid	methane
propanoic acid	ethane
	butane

Solution

It's tempting to glance at this question and just write down 'butanoic acid' as the next member of the series. Watch out – the alkanes have a different number of carbon atoms from the acid they're made from. Ethanoic acid (2 carbons) gives methane (1 carbon). That's not surprising when you think that carbon dioxide is given off in this reaction. The acids must have 1 carbon more than the alkanes. Butane has 4 carbon atoms, so the acid must have 5 carbons. That is **pentanoic acid**.

Example 8

Methylpropene and an alkane can be used to produce 2,2,4-trimethylpentane, a molecule added to petrol.

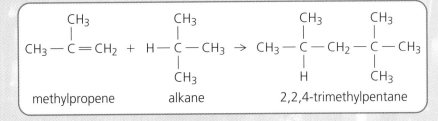

methylpropene   alkane   2,2,4-trimethylpentane

a) Give the systematic name for the alkane used to produce 2,2,4-trimethylpentane.

b) Name the type of chemical reaction shown above.

c) A similar reaction can be used to prepare 2,2-dimethylpentane.

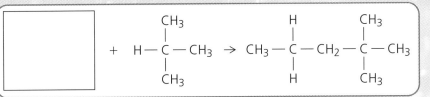

Draw a structural formula for the alkene used to form this molecule.

Solution

a) This isn't problem solving but it's still important to be able to answer it. The longest chain of carbon atoms is 3 – it doesn't matter which direction in the molecule you take. So it's a propane. It has a one carbon branch in the middle – so it's **methylpropane**. You don't have to put in a number, like 2-methylpropane, because there's no other position for a branch. If the fourth carbon atom was at the end of the molecule then it would be butane.

b) What's happened in this reaction? The alkene has lost its double bond. The alkane has been split like this:

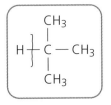

The H has been added to one side of the double bond, and the chunk on the right with the 4 carbons has been added to the other side of the double bond. It's an **addition reaction** – always a fair bet if a double bond vanishes.

c) Look to see how the two products are different. There is the first one:

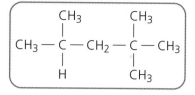

and the new one:

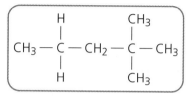

The molecules are very like each other, but the second carbon (from the alkene) doesn't have a methyl group on it. It has 2 hydrogen atoms. So instead of the alkene being:

$$CH_3 - \underset{\underset{CH_3}{|}}{\overset{\overset{CH_3}{|}}{C}} = CH_2$$

it must be like this:

$$CH_3 - \underset{\underset{H}{|}}{\overset{\overset{H}{|}}{C}} = CH_2$$

Example 9

Ethers are useful chemicals. Some ethers are listed in the table below.

Structural formula	Name of ether
$CH_3CH_2 - O - CH_2CH_3$	ethoxyethane
$CH_3 - O - CH_2CH_2CH_3$	methoxypropane
$CH_3 - O - CH_2CH_3$	methoxyethane
$CH_3CH_2 - O - CH_2CH_2CH_3$	X

a) Suggest a name for ether X.

b) The boiling points of ethers and alkanes are approximately the same when they have a similar relative formula mass. Suggest a value for the boiling point of ethoxyethane.

   (You may wish to use page 6 of the Data Booklet.)

Solution

a) If you look at the names, you can see that they're in three parts. The first bit tells you how many carbon atoms there are to the left of the oxygen. If it's 'eth' there are 2. If it's 'meth' there's only 1.

The second bit of the name, 'oxy', tells you that there's oxygen in the compound. The last bit of name tells you how many carbon atoms there are to the right of the oxygen. Molecule X has 2 carbons to the left of the oxygen, so it's 'ethoxy'. To the right of the oxygen, there are 3 carbons. So it's **ethoxypropane**.

b) You'll have to work out the formula mass of ethoxyethane first.

Next you have to decide which alkane has a formula mass close to 74. It's a fair bet that it will have at least 4 carbon atoms because ethoxyethane has 4. Give it another 1 to make up the mass of the oxygen. That's 5 carbon atoms – and you're at pentane, C_5H_{12}. Work out its relative formula mass – that's right, it's 72. Check its boiling point in the Data Booklet – it's **36 °C**.

$$CH_3\ CH_2 - O - CH_2\ CH_3$$
$$15 + 14 + 16 + 14 + 15 = 74$$

Example 10

The voltage of a cell is affected by the metals used as electrodes.

A student recorded the voltage and direction of electron flow for different cells.

Use the information in the table to predict the direction of electron flow and voltage you would expect when nickel and zinc electrodes are used.

Electrodes	Direction of electron flow	Voltage /V
Cu / Zn	Zn → Cu	1.0
Cu / Ni	Ni → Cu	0.5
Fe / Ni	Fe → Ni	0.2
Fe / Zn	Zn → Fe	0.3
Ni / Zn		

Solution

This question doesn't tell you to look at the ECS in the Data Booklet. If you do then it makes it a bit easier – it's quite difficult otherwise. If you set out the metals as they are in the Data Booklet and put in the voltages produced by pairs of metals you see something like this:

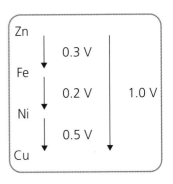

The arrows show the direction of electron flow. You can see that the voltages for each small step add up to the total for the big step. It looks as if the voltage when zinc is joined to nickel will be **0.5 V** and **electrons flow from zinc to nickel**.

You can do this in a slightly different way. Arrange the cells one after the other like this:

$$Zn \rightarrow Fe \ \ 0.3 \ V$$
$$Fe \rightarrow Ni \ \ 0.2 \ V$$
$$Ni \rightarrow Cu \ \ 0.5 \ V$$
$$Zn \rightarrow Cu \ \ 1.0 \ V$$

Again, you can see that the voltages from zinc to copper add up to the same whether it happens in one step or three. Again, for $Zn \rightarrow Ni$ you get 0.5 V with electrons flowing from zinc to nickel.

Flow diagram problems

Flow diagram problems appear frequently in Intermediate 2 papers. This is partly because it is possible to ask a number of questions based on the same diagram. Often, the questions are a mixture of the problem solving and knowledge and understanding types.

Example 11

Sodium carbonate is an important industrial chemical which is made from sodium chloride in the Solvay process – this is shown in Figure 5.2.

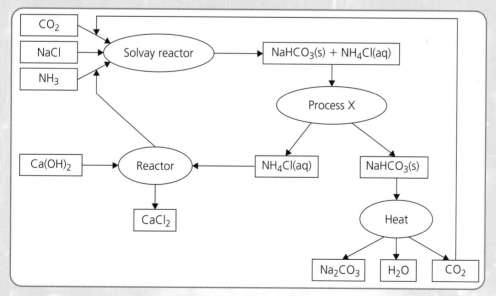

Figure 5.2 The Solvay process

a) Name process X.

b) Identify a substance which is recycled.

Example continued ➤

> ### Example 11 *continued*
>
> c) Sodium carbonate, Na_2CO_3, is the main product of the Solvay Process and is a salt. Name another salt produced.
>
> d) The main use for sodium carbonate is glassmaking for which a high purity is required. The purity of a sample of sodium carbonate can be checked by titration with acid.
>
> $$Na_2CO_3 + 2HCl \rightarrow 2NaCl + CO_2 + H_2O$$
>
> $22.4 \, cm^3$ of $0.1 \, mol \, l^{-1}$ HCl was required to neutralise $25 \, cm^3$ of a sodium carbonate solution.
>
> Calculate the concentration, in $mol \, l^{-1}$, of the sodium carbonate solution.

Solution

The first thing you should do when faced with a question like this is to inspect it closely, trying to understand what is going on, without even looking at the questions based on the diagram. Try to understand the chemistry – almost certainly, it will be simple. Besides, the chemical industry likes reactions which are simple, because they are usually economical to carry out.

The problem is – where to start? The key to the process is the Solvay reactor. In go three substances – carbon dioxide, sodium chloride and ammonia. Out come ammonium chloride and sodium hydrogencarbonate. Somehow, they are separated. The sodium hydrogencarbonate is heated and breaks down to sodium carbonate (the end product), carbon dioxide and water. The ammonium chloride is made to react with calcium hydroxide. The products are calcium chloride and ammonia. The ammonia is passed back into the reaction.

a) Process X separates a mixture of solid sodium hydrogencarbonate and ammonium chloride solution. To separate a solid from a solution, it is filtered. **Filtration** must be the answer to this question.

b) If you inspect the diagram you'll see a big arrow carrying carbon dioxide, produced as an end product, back to the very start of the process. You'll also see that ammonia, formed by the reaction of ammonium chloride and calcium hydroxide, is fed back to the start of the process. So you can answer either **carbon dioxide** or **ammonia** (or both!).

c) You have to find a salt produced. What's a salt? It's a compound which contains a metal ion (or an ammonium ion) and an ion from an acid. There aren't too many other products to choose from – water, carbon dioxide (which is recycled) and calcium chloride. Calcium is a metal and the chloride ion comes from hydrochloric acid. **Calcium chloride** is the answer.

d) This isn't problem solving but it's important that you're confident with calculations of this type. If you look back at Chapter 3, you can remind yourself of ways of doing this. What follows is probably the safest way to go about it.

First calculate the number of moles of acid used. That's the volume in litres multiplied by the concentration in moles per litre:

$$\text{moles of acid} = 0.0224 \times 0.1 = 0.00224 \text{ mole}$$

If you look at the equation, you'll see that 2 moles of acid react with 1 mole of sodium carbonate. That means that there are half as many moles of sodium carbonate as there are moles of acid:

$$\text{moles of sodium carbonate} = \frac{0.00224}{2} = 0.00112$$

This number of moles is contained in 0.025 litre of sodium carbonate solution.

The concentration in moles per litre is:

$$\frac{\text{number of moles}}{\text{number of litres}} = \frac{0.00112}{0.025} = \textbf{0.0448 mol l}^{-1}$$

Even if you can't see your way to the end of the calculation, do what you can. If you're told the volume of a solution and its concentration, always work out the number of moles present. That's nearly always worth half a mark. If you copy out the equation and write the number of moles under the relevant chemicals, like this:

$$Na_2CO_3 \;+\; 2HCl \;\rightarrow\; 2NaCl \;+\; CO_2 \;+\; H_2O$$

1 mole   2 moles

… you'll almost certainly get half a mark. And maybe, by the time you've done these things, you'll see how to complete the calculation!

Example 12

Propane can be cracked to give a mixture of smaller molecules – this process is shown in Figure 5.3.

a) Catalysts can be used to speed up chemical reactions. Give another advantage of using a catalyst.

b) Name the process used in Step 2 to separate the product mixture.

c) Complete the flowchart by naming the other product separated from the mixture.

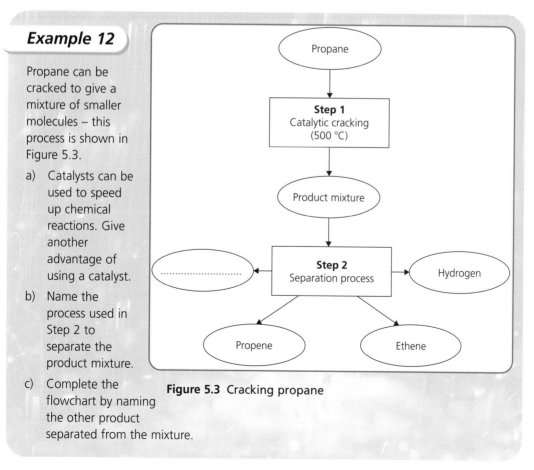

Figure 5.3 Cracking propane

Solution

a) This is just knowledge and understanding. The usual answer is that it results in **less energy being needed**. You'll notice that a temperature of 500 °C is used. If the catalyst wasn't there then a higher temperature would be needed. This would make the process more expensive. Another possible answer is that the catalyst can be reused. You should be able to say what the catalyst is – it's aluminium oxide.

b) The products are all gases at room temperature. In industry these would be cooled to form a liquid mixture. All the components of the mixture have different boiling points, so they can be separated by **fractional distillation**. Fractional distillation is pretty well the only separation process used in the oil industry, so it's always a fair bet that this will be the answer to a question which asks what process is used to separate hydrocarbons!

c) This is where you need to think. Cracking breaks alkane molecules down into smaller hydrocarbons. A mixture of alkanes and alkenes is produced because there aren't enough hydrogen atoms present to form alkanes only. The possible molecules from propane are propene (with 3 carbon atoms), ethene and ethane (with 2) and methane (with 1). And, of course, hydrogen. You can't pick hydrogen, ethene or propene because they are already in the flowchart. So it's methane or ethane.

Think about ethane for a moment. Propane is C_3H_8 and ethane is C_2H_6. If you cracked propane to give ethane, the balanced equation would be:

$$C_3H_8 \rightarrow C_2H_6 + CH_2$$

But there's no such molecule as CH_2. Carbon's valency is 4, not 2! So it looks as though you'll have to pick methane as the product. If methane forms from the cracking of propane the balanced equation would be:

$$C_3H_8 \rightarrow CH_4 + C_2H_4$$

These are both real compounds. **Methane** must be the answer.

Example 13

Sulphuric acid is an important industrial chemical. It can be made by oxidising ammonia to produce nitrogen monoxide. Water is also produced in the reaction. The nitrogen monoxide is reacted with more oxygen to form nitrogen dioxide. The sulphuric acid can then be produced by reacting the nitrogen dioxide with sulphurous acid.

a) Complete the flowchart in Figure 5.4 to show this process.

b) When nitrogen dioxide reacts with sulphurous acid to produce sulphuric acid, nitrogen monoxide is also be produced. This can be recycled.

Add a line to the flowchart to show nitrogen monoxide being recycled.

c) Sulphuric acid is a strong acid and sulphurous acid is a weak acid. Why is sulphuric acid described as a strong acid?

Example continued ➤

Example 13 *continued*

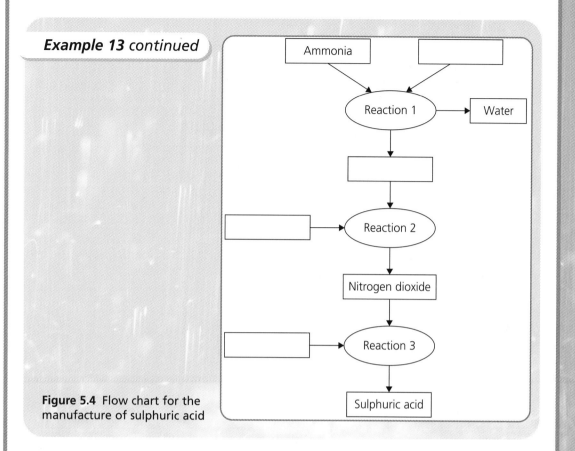

Figure 5.4 Flow chart for the manufacture of sulphuric acid

Solution

a) It's just a matter of reading the question carefully and putting the correct chemicals in the boxes. From top to bottom the boxes should read **oxygen** (or 'air'), **nitrogen monoxide**, **oxygen** and **sulphurous acid**.

Just make sure you take care not to make slips like writing 'ammonium' in the top box, or 'sulphuric acid' in the last box because mistakes like that will cost you half a mark each.

HOW TO PASS INTERMEDIATE 2 CHEMISTRY

b)   The arrow is shown in bold in the diagram:

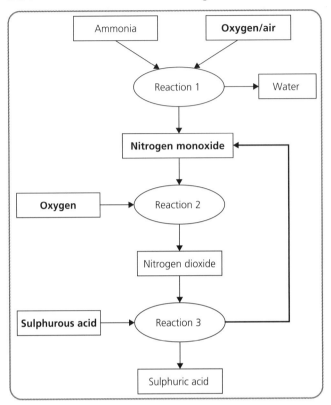

c)   This is just knowledge and understanding. A strong acid is an acid that **breaks up fully into ions when in water** – it is fully ionised or fully dissociated.

Unfamiliar settings

Some questions are of the problem-solving type because they present information in an unfamiliar setting, or present you with new information and ask you to draw conclusions from it. Here are a few examples of questions like this.

Example 14

When dilute hydrochloric acid is added to substance X, a gas is given off. This gas puts out the candle flame quickly.

Which of the following could be substance X?

A   Calcium hydroxide

B   Calcium carbonate

C   Calcium oxide

D   Calcium

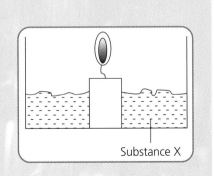

Solution

It shouldn't be a matter of guessing! Your revision should tell you that dilute hydrochloric acid will react with all four of these substances. In A, the reaction will produce calcium chloride and water, as will the reaction in C. The equations for these reactions are:

$$Ca(OH)_2 + 2HCl \rightarrow CaCl_2 + 2H_2O$$

$$CaO + 2HCl \rightarrow CaCl_2 + H_2O$$

In B, the reaction produces carbon dioxide – remember, any acid added to any carbonate produces carbon dioxide. This is used as a test to show that a substance is a carbonate. The equation for the reaction is:

$$CaCO_3 + 2HCl \rightarrow CaCl_2 + H_2O + CO_2$$

This is the one you want – carbon dioxide will put out the candle – after all, it's used in fire extinguishers, isn't it? **B** is the answer.

Don't be tempted to try the experiment with D, calcium. Calcium will react rapidly with acid to produce hydrogen. And hydrogen is an explosive gas!

Example 15

Dinitrogen monoxide can be used to boost the performance of racing car engines.

When dinitrogen monoxide decomposes it forms a mixture of nitrogen and oxygen. The reaction is exothermic.

$$2N_2O \rightarrow 2N_2 + O_2$$

a) What is meant by an exothermic reaction?

b) How many moles of oxygen will be produced when 4 moles of dinitrogen monoxide are decomposed?

c) An experiment with two identical candles was set up as shown below.

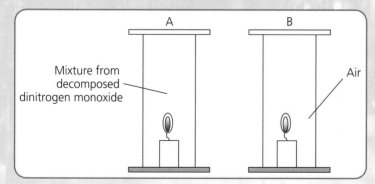

Why will the candle burn for longer in gas jar A?

Solution

a) This is just a bit of knowledge and understanding. An exothermic reaction is one which gives out heat.

b) The equation for the reaction is:

$$2N_2O \rightarrow 2N_2 + O_2$$

which tells us that 2 moles of dinitrogen monoxide produce 1 mole of oxygen. It follows that 4 moles of the gas must produce **2 moles** of oxygen.

c) This is the problem-solving part. Why does the candle in jar A burn longer? To answer this you'll need to know something about how the air is made up. The air is mainly nitrogen – about 78% or so. About 20% of the air is oxygen, and there's some argon and various other gases. The key point, for this question, is that 20% of the air is oxygen. The mixture of gases in jar A contains 2 moles of nitrogen to 1 mole of oxygen. That is, in every 3 moles of gas, 1 mole is oxygen. This suggests that about one-third of the gas in jar A is oxygen. This is more than in air, where only about one-fifth is oxygen. The candle burns longer, because **there's more oxygen for it to burn in**.

Example 16

A membrane cell, shown in Figure 5.5, can be used to produce chlorine, hydrogen and sodium hydroxide from sodium chloride solution.

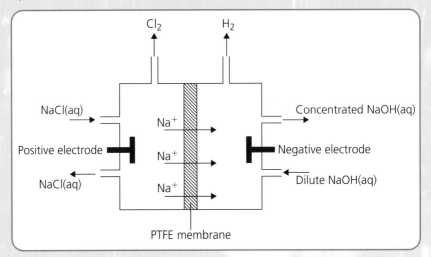

Figure 5.5 The electrolysis of sodium chloride solution

The ion–electron equations for the electrode reactions are:

positive electrode: $2Cl^- \rightarrow Cl_2 + 2e^-$

negative electrode: $2H_2O + 2e^- \rightarrow 2OH^- + H_2$

a) Combine the ion–electron equations to produce a redox equation.

Example continued ➤

Example 16 continued

b) The membrane is made up of layers of poly(tetrafluoroethene) (PTFE) with negatively charged groups attached. Sodium ions pass through the membrane towards the negative electrode.

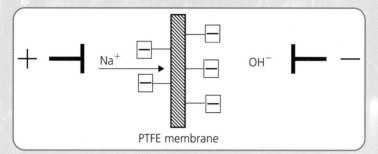

PTFE membrane

Suggest one reason why hydroxide ions (OH^-) will not pass through the membrane toward the positive electrode.

c) PTFE can be reshaped on heating. What name is given to plastics which can be reshaped on heating?

Solution

Only part (b) of this question is problem solving, but the complicated diagrams might make you think the whole thing is difficult! It's actually not too hard. You just have to read it carefully and study the diagrams.

a) This isn't too bad. You combine ion–electron equations by cancelling out equal numbers of electrons on both sides of the arrows, and adding together what's left.

$$2Cl^- \rightarrow Cl_2 + 2e^-$$
$$2H_2O + 2e^- \rightarrow 2OH^- + H_2$$
$$\overline{2H_2O + 2Cl^- \rightarrow 2OH^- + Cl_2 + H_2}$$

If the equations hadn't contained equal numbers of electrons on both sides, you would have had to multiply one or both equations to get the numbers of electrons equal.

b) This is the problem-solving bit of the question. It's quite easy – hydroxide ions have a negative charge and so has the surface of the PTFE membrane. **Similar charges repel each other**. The hydroxide ions won't be able to get near the membrane, far less pass through it!

c) This is a simple bit of knowledge and understanding. Polymers that can be reshaped on heating are called '**thermoplastic**'. The other type is 'thermosetting'. This term is applied to polymers which become hard when they are heated, and can't be melted by further heating. You just have to learn these words and their meanings.

Example 17

Sodium carbonate reacts with dilute hydrochloric acid.

a) Write the formula for sodium carbonate.

b) Name the salt formed when sodium carbonate reacts with hydrochloric acid.

c) A teacher used this reaction in an experiment to show that carbon dioxide puts out a flame.

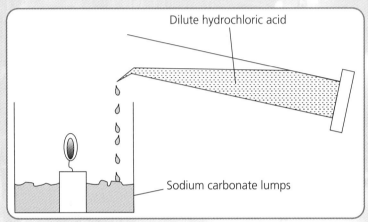

Dilute hydrochloric acid

Sodium carbonate lumps

Why would the candle burn for longer if an equal volume and concentration of ethanoic acid was used instead of the dilute hydrochloric acid?

Solution

a) This is knowledge and understanding. You should be able to write chemical formulae easily. One way of doing this is to check what group sodium is in. It's in Group 1 of the Periodic Table – this tells us its valency is 1. You should know that the formula for the carbonate ion is CO_3^{2-}. If you can't remember this, you'll find it in the Data Booklet on page 4. Because its charge is 2–, its valency is 2. By 'swapping valencies' you find that the formula is **Na_2CO_3**.

b) It shouldn't be too difficult to name the salt. Salts contain an ion from a metal and an ion from an acid. There's only one metal in this reaction – sodium. The acid is hydrochloric acid, so the ion from the acid is chloride. So the salt is **sodium chloride**.

Watch out! If you write the formula, instead of giving the name, make sure you get the formula correct – NaCl. If you give a wrong formula then you'll lose the mark.

(c) Now for the problem-solving part. The answer must be something to do with the fact that it's ethanoic acid being used instead of hydrochloric acid. The main difference between these acids is that hydrochloric acid is a strong acid and ethanoic acid is a weak acid. The strong acid is split fully into ions. The weak acid is only partly split into ions. If you think about the reaction between carbonates and acids, the key part of the reaction is this:

$$2H^+ + CO_3^{2-} \rightarrow H_2O + CO_2$$
from   from
acid   carbonate

Because the weak acid contains only a few H^+ ions, the rate of reaction between the acid and the sodium carbonate will be slower than with hydrochloric acid with lots of H^+ ions. If the reaction is slower, **carbon dioxide will be released more slowly** so the candle will burn longer.

Example 18

Scientists have developed self-heating food packs. They use the heat given out by the reaction of magnesium with water. The magnesium is in the form of an alloy with iron.

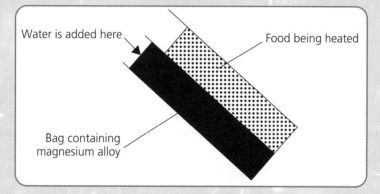

Water is added here

Food being heated

Bag containing
magnesium alloy

a) What term is used to describe reactions which give out heat?

b) The reaction is started by the addition of water to the bag containing magnesium alloy. The equation for the reaction is:

$$Mg + 2H_2O \rightarrow Mg(OH)_2 + H_2$$

Why is it necessary to keep the food bag away from flames when the food is being heated?

c) In this reaction, magnesium atoms lose electrons:

$$Mg \rightarrow Mg^{2+} + 2e^-$$

What name is given to this type of chemical reaction?

d) In the alloy, magnesium is in contact with iron. This contact speeds up the reaction and produces heat more quickly. Suggest why the magnesium being in contact with iron speeds up the reaction.

Solution

a) This question again! It's **exothermic** – obviously a word worth knowing for exams!

b) You might think that it would be a good idea to have the food bag near flames if you were trying to heat the food! However, you'll notice that one of the products of the reaction is the gas hydrogen. **Hydrogen is flammable** – explosive when mixed with air in the right proportion. So it's a good idea to keep the bag away from flames!

c) All you have to remember is that loss of electrons is oxidation and gain of electrons is reduction. Remember OIL RIG – **o**xidation **is l**oss, **r**eduction **is g**ain'? So this reaction is an example of **oxidation**.

d) It's difficult to see the answer to this problem. One of the acceptable answers is that the **iron acts as a catalyst**. Another acceptable answer is that because iron is below magnesium in the ECS then electrons flow from magnesium to iron, and that somehow speeds up the reaction.

Example 19

Four cells were made by joining copper, iron, silver and tin in turn to zinc.

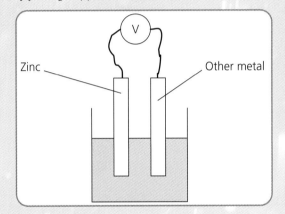

The voltages obtained are shown in the table.

Cell	Voltage /V
A	1.5
B	1.1
C	0.6
D	0.3

Which line in the table shows the voltage of the cell containing iron joined to zinc?

(You can use page 7 of the Data Booklet to help you.)

Solution

If you look at the Data Booklet, you see that the positions of the five metals in the ECS are as follows:

> zinc
>
> iron
>
> tin
>
> copper
>
> silver

The closer two metals are together in the ECS, the more similar is their ability to attract electrons. This means that the voltage of a cell in which they are involved will be low. Because iron is closest to zinc in the ECS, the iron / zinc cell will have the lowest voltage, giving **D** as the answer.

PRESCRIBED PRACTICAL ACTIVITIES

Section B of your exam contains about 5 marks in questions on PPAs. Usually there are two questions on PPAs and they are flagged up. The questions say that they are about PPAs and tell you (in bold print!) which PPA it is. For example, '**In the Unit 1 PPA 'The effect of concentration on reaction rate** ...'.

The marks are easy to get, if you learn your stuff!

The exam tests purely practical aspects of the PPAs – not theoretical detail. You must know the practical details thoroughly – what chemicals are used, what was done, what measurements or observations were made and what safety precautions should be taken. You'll usually find that a question asking about a PPA also asks about other things as well, related to the PPA but not about the actual PPA itself.

Here's a list of the PPAs.

Unit 1:
- The effect of concentration changes on reaction rate
- The effect of temperature changes on reaction rate
- Electrolysis

Unit 2:
- Testing for unsaturation
- Cracking
- Hydrolysis of starch

Unit 3:
- Preparation of a salt
- Factors affecting voltage
- Reaction of metals with oxygen

We'll summarise the PPAs for each unit in turn, and then look at some examples for each unit.

Unit 1

Effect of concentration on reaction rate

This experiment involves the reaction of sodium persulphate solution with potassium iodide solution in the presence of starch. The potassium iodide solution also includes a small quantity of sodium thiosulphate.

The reaction releases iodine which reacts at once with the sodium thiosulphate. When all the sodium thiosulphate has been used up by the iodine, a surplus of iodine gathers in the beaker and reacts with the starch turning it blue/black.

In this experiment, the concentration of sodium persulphate is altered as follows:

Volume of sodium persulphate /cm^3	Volume of water /cm^3
10	0
8	2
6	4
4	6

Each experiment is done with 10 cm^3 of potassium iodide solution and 1 cm^3 of starch solution. Deionised water is used to prevent the addition of impurities which might affect the rate of the reaction.

10 cm^3 of sodium persulphate is placed in a beaker and 1 cm^3 of starch is added. The beaker is placed on a white tile to make it easier to spot the end of the reaction. 10 cm^3 of potassium iodide solution is added and a timer started. The timing stops when the mixture suddenly goes blue/black.

The experiment is repeated with the three mixtures in the table above. Making the volume up to 10 cm^3 each time with water ensures that no other concentration alters.

Effect of temperature on reaction rate

This experiment uses the reaction between sodium thiosulphate solution and dilute hydrochloric acid. When these solutions are mixed, a precipitate of sulphur gradually forms – this makes the mixture go cloudy. If the beaker in which the reaction is taking place is on a piece of paper marked with a cross, the cross will eventually become invisible to someone watching from a position over the beaker. Sulphur dioxide gas is given off in this reaction and suitable precautions, such as adequate ventilation, are required.

The first experiment is carried out at room temperature. 20 cm^3 of sodium thiosulphate solution is placed in a 100 cm^3 beaker sitting on a piece of paper marked with a cross. 1 cm^3 of hydrochloric acid is added and a timer started. The timing stops when the cross is no longer visible. The initial temperature of the mixture is taken as the reaction temperature.

The 100 cm^3 beaker is now cleaned thoroughly before being used again – including removing any sulphur which might interfere with the visibility of the cross.

The beaker in which the sodium thiosulphate solution is stored is then heated to about 30 °C and the experiment is repeated. The reaction temperature is taken as the temperature of the solutions on mixing.

The experiment is then repeated at about 40 °C and 50 °C.

Electrolysis

In this experiment, a solution of copper(II) chloride is electrolysed using carbon electrodes. Copper forms at the negative electrode as copper ions gain electrons:

$$Cu^{2+}(aq) + 2e^- \rightarrow Cu(s)$$

Chlorine forms at the positive electrode as chloride ions lose electrons:

$$2Cl^-(aq) \rightarrow Cl_2(g) + 2e^-$$

Bubbles of gas can be seen rising from the positive electrode. A piece of moist blue litmus paper held over the positive electrode turns red and is then bleached.

The copper can be detected as a red/brown coating on the black carbon rod. The chlorine is detected by gently wafting the hand over the electrolysis cell towards your nose (care!). The smell is like the smell of bleach.

Example 1

In the PPA 'Effect of concentration on reaction rate', the reaction between sodium persulphate and potassium iodide solutions is studied The apparatus used is shown in Figure 6.1.

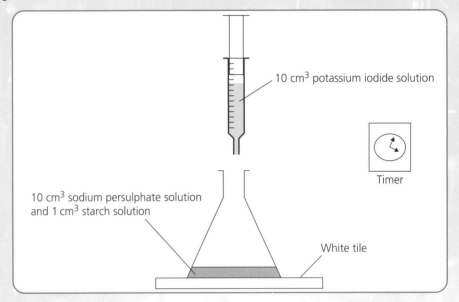

10 cm³ potassium iodide solution

Timer

10 cm³ sodium persulphate solution and 1 cm³ starch solution

White tile

Figure 6.1 The effect of concentration on reaction rate

Results

Experiment	1	2	3	4
Volume of sodium persulphate solution /cm³	10	8	6	4
Volume of water /cm³	0			
Reaction time /s	126	162	210	336

a) Complete the table to show the volumes of water used in experiments 2, 3 and 4.

b) How would you know when to stop the timer in each experiment?

c) Why are volumes of sodium persulphate solution less than 4 cm³ not used?

d) The formula for the persulphate ion is $S_2O_8^{2-}$. Write the formula for sodium persulphate.

Solution

a) If you understand that it's necessary to keep the total volume constant (otherwise all the concentrations will change, not just that of sodium persulphate) then you should have been able to complete the table like this:

Experiment	1	2	3	4
Volume of sodium persulphate solution /cm³	10	8	6	4
Volume of water /cm³	0	2	4	6
Reaction time /s	126	162	210	336

b) You might have worked out the answer to (a) even if you'd never done the experiment. But you must **know** the answer to (b) – you can't work it out. After a time, iodine accumulates in the mixture and the **contents of the beaker go blue/black** in a very short space of time. That's when you stop timing.

c) As you can see, the less sodium persulphate solution used, the slower the reaction. If less than 4 cm³ of the solution is used then the reaction will be so slow that the colour change at the end will be gradual, and **you won't be absolutely sure when to stop timing**.

d) This isn't a PPA question at all but you should be able to do it. The charge of the persulphate ion is 2– so its valency is 2. Sodium is in Group 1 so its valency is 1. Swapping the valencies gives the formula **$Na_2S_2O_8$**.

Example 2

Sodium thiosulphate reacts with hydrochloric acid to produce a precipitate of sulphur. The equation for the reaction is:

$$Na_2S_2O_3 + 2HCl \rightarrow 2NaCl + S + SO_2 + H_2O$$

This reaction is used in a PPA to study the effect of temperature on reaction rate. The rate is determined from the time taken to produce a certain amount of sulphur.

a) How would you decide when to stop timing?

b) The experiment must be carried out in a well ventilated area. Give a reason for this.

c) A student investigated the effect of changing the concentration of the sodium thiosulphate solution on the reaction rate. The results obtained are shown in the graph.

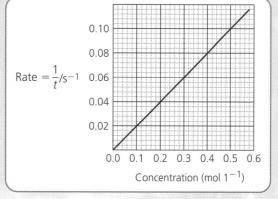

Use the graph to find the time taken, in seconds, when the experiment was carried out using 0.1 mol l⁻¹ sodium thiosulphate solution.

Solution

a) If you remember, this is the reaction where the beaker sits on a piece of paper marked with a cross. You wait for the precipitate of sulphur to get so thick that **you can't see the cross**.

b) You'll notice that one of the products of the reaction is the gas sulphur dioxide. This is **toxic and can induce attacks of asthma in some people**. That's why ventilation is needed.

c) Watch out! You're not being told just to read the graph where the concentration of sodium thiosulphate is 0.1 mol l^{-1}. This is because it's a time you're to find, not a rate. The rate at this point is 0.02 s^{-1}. This is equal to $1/t$, so:

$$t = \frac{1}{0.02} = 50$$

So **50 seconds** is the answer.

Example 3

In the PPA 'Effect of temperature changes on reaction rate', the rate of reaction between sodium thiosulphate and hydrochloric acid was investigated using the apparatus in Figure 6.2.

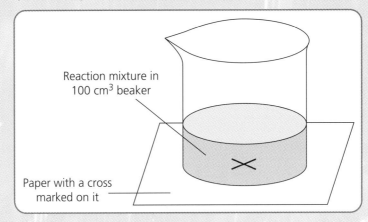

Reaction mixture in 100 cm³ beaker

Paper with a cross marked on it

Figure 6.2 The effect of temperature on reaction rate

a) Describe the change that would take place in the beaker.

b) When the experiment was carried out at the different temperatures, either the same beaker or identical beakers had to be used. Why is it important to use the same or identical beakers in this experiment?

Solution

a) You know the answer to this – a precipitate will form so it will **become cloudy**.

b) The experiments should all involve the same mixture of sodium thiosulphate and acid. The volume of liquid is the same in each experiment. **The only thing that's different should be the temperature**. Suppose you did one experiment in a 100 cm³ beaker and a second experiment in a larger 250 cm³ beaker. In the big beaker, there would be a shallower depth of liquid. That would affect the time taken for the cross to be hidden, because there would be less sulphur precipitate through which to look at the cross.

Example 4

In the PPA 'Electrolysis', copper(II) chloride is separated into its elements. Figure 6.3 shows the apparatus used.

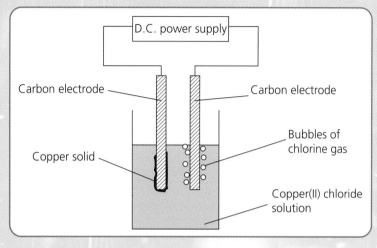

Figure 6.3 Electrolysing a solution of a copper(II) salt

a)   Label the diagram to show the charge on each electrode.

b)   Describe how to smell chlorine gas safely.

c)   During the electrolysis, copper ions are changed to copper atoms. Why is this reaction described as reduction?

d)   By the end of this experiment, 2.10 g of copper was deposited. Calculate the number of moles of copper deposited.

Solution

a)   One electrode must be positive and the other must be negative – that's because a direct current (d.c.) is being used. If an a.c. current was used then the charge on each electrode would change 50 times a second and no electrolysis would take place.

   Metals generally form positive ions because they have just a few electrons in their outer shell and become stable by losing them. This leaves the metal atom as a positively charged metal ion. Because opposite charges attract, the **left-hand electrode must be negatively charged** because the copper is forming there, and copper ions are positively charged. That means the **right-hand electrode must be positively charged**. Chlorine is forming there and chloride ions have a negative charge.

b)   This is what the PPA instructions say: 'To smell the gas given off at the positive electrode, first breathe in deeply to fill the lungs with uncontaminated air. With your nose at least 30 cm from the electrolytic cell gently waft your hand over the cell towards your nose. Take just a sniff of the gas'. However, if you say something like '**waft the gas in the direction of the nose**' or '**smell it from a distance**' you should get the mark.

c)   This isn't a PPA question, but it's important to be able to answer it. For copper ions to turn into copper atoms, they have to gain electrons. The ion–electron equation for the process is:

$$Cu^{2+} + 2e^- \rightarrow Cu$$

This is a reduction reaction because **the gain of electrons is reduction** (OIL RIG).

d) 2.10 g of copper was deposited. To convert this to moles, you have to divide the mass by the mass of one mole of copper (the relative atomic mass of copper expressed in grams):

$$\frac{\text{mass of copper made}}{\text{mass of 1 mole of copper}} = \frac{2.10}{63.5} = 0.033 \text{ mole}$$

Unit 2

Testing for unsaturation

In this PPA, four hydrocarbons (A, B, C and D) are tested with bromine solution. The hydrocarbons have formulae as follows:

◆ A is C_6H_{14}

◆ B and C are C_6H_{12}

◆ D is C_6H_{10}.

A 0.5 cm depth of a different hydrocarbon is added to each of four test tubes. About ten drops of bromine solution are added to each tube. The tubes are shaken gently and the observations – whether or not the bromine is decolourised – are noted. From the observations, the hydrocarbons can be classified as saturated or unsaturated.

The main safety point in this PPA is to make sure that there are no flames about – all the hydrocarbons are flammable. Because bromine is corrosive, sodium thiosulphate is used to counteract its effects if it is spilled onto the skin.

Cracking

Cracking is used to break long chain hydrocarbon molecules down into shorter ones.

The apparatus shown in Figure 6.4 is set up.

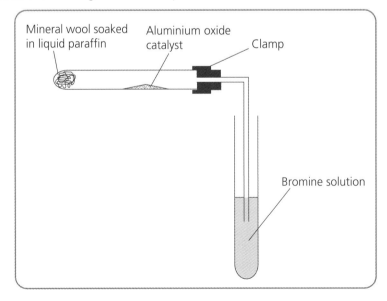

Mineral wool soaked in liquid paraffin

Aluminium oxide catalyst

Clamp

Bromine solution

Figure 6.4 Cracking hydrocarbons in the laboratory

The aluminium oxide catalyst is heated strongly for some time. The heat is then transferred to the mineral wool to vaporise some of the liquid paraffin. The vapour from the paraffin passes over the hot catalyst and cracking takes place. The heat is moved from the mineral wool to the catalyst, then back to the mineral wool, and so on – the aim is to keep both fairly hot. Bubbles are seen coming from the end of the delivery tube into the bromine solution, and the bromine is eventually decolourised.

At the end of the experiment, the clamp stand must be lifted to remove the delivery tube completely from the bromine solution, otherwise 'suckback' will occur. The gases in the heated tube will cool down and contract as they cool. This will result in cold bromine solution being sucked back into the heated tube, perhaps shattering it when the cold liquid meets the hot glass.

Figure 6.5 Suckback can go with a bang!

Hydrolysis of starch

Starch, which is a condensation polymer of glucose, can be broken down into smaller sugar molecules by the action of the enzyme amylase or by dilute hydrochloric acid. The presence of the smaller sugar molecules can be shown using Benedict's solution.

Enzyme method

3 cm^3 of starch solution is added to each of two test tubes. To one tube, 1 cm^3 of amylase solution is added. To the other, 1 cm^3 of water is added as a control experiment. Both tubes are placed in a warm water bath with a maximum temperature of 40 °C for 5 minutes. Then 2 cm^3 of Benedict's solution is added to both tubes. The temperature of the water in the water bath is then raised to boiling point. The tube which contained the amylase should give a positive result with Benedict's solution showing that the enzyme has hydrolysed it.

Acid option

10 cm^3 of starch solution is added to each of two small beakers. To one beaker, 1 cm^3 of dilute hydrochloric acid is added. The same volume of water is added to the other beaker as a control experiment. The beakers are then heated gently on a tripod stand over a Bunsen burner until they boil. They are allowed to boil for 5 minutes, and then to cool a little. Very small quantities of sodium hydrogencarbonate are added to the acid mixture until no more bubbles of carbon dioxide are given off. This is to neutralise the acid because Benedict's solution does not work in acidic solutions. 5 cm^3 of Benedict's solution is now added to each beaker and the beakers are heated again. The tube which contained the acid should give a positive result for Benedict's solution showing that the acid has hydrolysed it.

Example 5

A student's results are shown below for the PPA 'Testing for unsaturation'.

Hydrocarbon	Molecular formula	Observation with bromine solution	Saturated or Unsaturated
A	C_6H_{14}	No change	Saturated
B	C_6H_{12}	Bromine decolourised	Unsaturated
C	C_6H_{12}	No change	Saturated
D	C_6H_{10}	Bromine decolourised	Unsaturated

a) Draw a possible structural formula for hydrocarbon D.

b) Sodium thiosulphate solution is made available as a safety measure, because one of the chemicals used in the experiment is corrosive. Which chemical is corrosive?

Solution

a) The chances are that hydrocarbon D is cyclohexene, with the following formula:

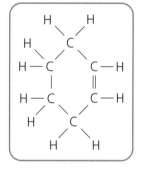

But there's nothing to stop you suggesting

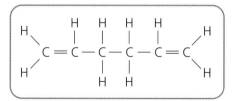

or some other alkene with 2 carbon to carbon double bonds. Just make sure that every carbon atom forms a total of 4 bonds – a double bond counts as 2 – and that every hydrogen atom forms 1 bond.

b) This is just a simple case of remembering that **bromine solution** is corrosive.

Example 6

The diagram below shows the apparatus used in the PPA 'Cracking'. Liquid paraffin is cracked using an aluminium oxide catalyst. Bromine solution is used to show that some of the products are unsaturated.

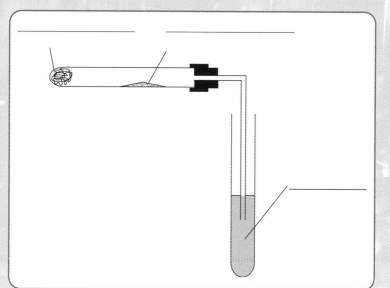

a) Label the diagram of the apparatus used to crack liquid paraffin.

b) What safety precaution should be taken before heating is stopped?

Solution

a) Look at the diagram under the heading 'Cracking' for the answers to this! Ignore the clamp!

b) At the end of the experiment the **heated tube must be lifted to remove the delivery tube completely from the bromine solution**, otherwise 'suckback' will occur. This is usually done by lifting the clamp stand which is supporting the heated tube.

Example 7

In the PPA 'Hydrolysis of starch', dilute hydrochloric acid can be used to break down the starch. Figure 6.6 shows suitable apparatus for testing the products.

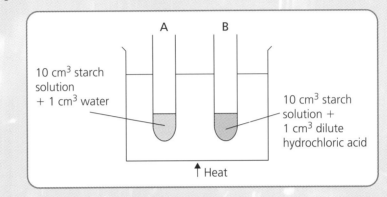

10 cm³ starch
solution
+ 1 cm³ water

10 cm³ starch
solution +
1 cm³ dilute
hydrochloric acid

↑ Heat

Figure 6.6 Hydrolysing starch

a) After heating with dilute hydrochloric acid solid sodium hydrogencarbonate is added to each reaction mixture. Why is sodium hydrogencarbonate added at this stage?

b) Complete the table to show the results which should be obtained when the reaction mixtures are tested with Benedict's solution.

Reaction mixture	Observation on heating with Benedict's solution
A	
B	

Solution

a) This is **to neutralise the acid** because Benedict's solution does not work in an acidic solution.

b) You can see that these PPA questions are very simple indeed if you learn your stuff!

Reaction mixture	Observation on heating with Benedict's solution
A	**No change – blue colour remains**
B	**Becomes orange/red**

Unit 3

Preparation of a salt

In this PPA, magnesium sulphate is prepared. It can be made by the reaction of dilute sulphuric acid with either magnesium metal, in the form of magnesium turnings, or with magnesium carbonate. Excess solid has to be added to the acid to make sure that all the acid has reacted and that neutralisation is complete. The advantage of using magnesium metal or magnesium carbonate is that you can tell when the reaction is complete because no more gas is given off.

The equations for the reactions are:

$$Mg + H_2SO_4 \rightarrow MgSO_4 + H_2$$

$$MgCO_3 + H_2SO_4 \rightarrow MgSO_4 + H_2O + CO_2$$

20 cm^3 of dilute sulphuric acid is placed in a beaker. A spatula of magnesium or magnesium carbonate is added with stirring. When all the solid has reacted, another spatulaful is added with stirring, and so on until there are no more bubbles of gas and some unreacted solid remains at the bottom of the beaker.

The unreacted solid is removed by filtration. The clear solution of magnesium sulphate is transferred to an evaporating basin and heated gently until about half the water has boiled away. It is then left to cool. White crystals of magnesium sulphate slowly form.

Factors affecting voltage

In this PPA, you investigate factors which might affect the voltage of a simple cell. The cell is of the type shown in Figure 6.7.

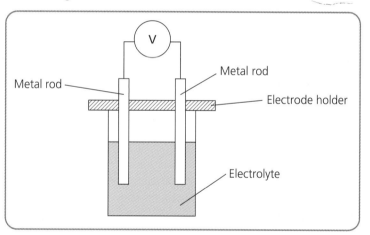

Figure 6.7 Investigating the voltage of simple cells

You can investigate either the metals used for the electrodes or the type of electrolyte.

Metals

You are supplied with rods of copper, zinc and iron. The copper and zinc rods have to be sanded with emery paper and washed before each use. The electrolyte is sodium chloride solution. You measure the voltage of the copper/zinc cell, the copper/iron cell and the zinc/iron cell. You're expected to take duplicate results by removing the rods from the solution, cleaning them as above, and then reconnecting them to the voltmeter.

Electrolyte

You are supplied with three solutions – sodium chloride, sodium hydroxide and hydrochloric acid. The electrodes are zinc and copper, which have to be cleaned before each use as above. Again, you're expected to take duplicate results as described above.

Reaction of metals with oxygen

This experiment lets you place zinc, copper and magnesium in order of reactivity by comparing the way in which they combine with oxygen. The following apparatus is used – the purpose of the potassium permanganate is to release oxygen when it is heated.

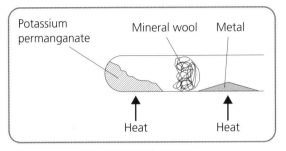

You have to make sure that the mouth of the test tube is not pointing at anyone. It's important to heat the metal strongly first, then move the heat to the potassium permanganate to generate oxygen and allow it to pass over the heated metal. When you're carrying out the experiment with magnesium, you should shade your eyes with your free hand and not look directly at the magnesium.

Example 8

Part of a student's PPA instruction sheet for 'Preparation of a salt' is shown below.

Preparation of a Salt

Aim
The aim of this experiment is to make a magnesium salt by the reaction of magnesium/magnesium carbonate with sulphuric acid.

Procedure

1 Using a measuring cylinder add 20 cm^3 of dilute acid to the beaker.

2 Add a spatula of magnesium (or magnesium carbonate) to the acid and stir the reaction mixture with a glass rod.

3 If all of the solid reacts, add another spatulaful of magnesium (or magnesium carbonate) and stir the mixture.

4 Continue adding the magnesium (or magnesium carbonate) until …

a) Complete instruction 4 of the procedure.

b) Why is an excess of magnesium or magnesium carbonate added to the acid?

c) There are three steps in the preparation of magnesium sulphate from magnesium or magnesium carbonate. Instructions 1 to 4, shown above, describe the 'reaction step'.

Name the next two steps.

Solution

a) If you read the introduction to 'preparation of a salt' you'll see that the advantage of using magnesium metal or magnesium carbonate is that a gas is released, so you can tell when you've added enough when the release of gas stops. So the answer is '**no more gas is given off**'.

b) An excess is added to make sure that **all the acid is neutralised**.

c) Once the neutralisation is complete the excess solid is filtered out and the solution is heated to evaporate about half of the liquid. So the answers are **filtration** and **evaporation**.

Example 9

In a PPA, students were asked to investigate how different metals affect the size of the voltage generated by a simple cell.

Their results are shown in the table.

Metals used	Average voltage /V
Iron and copper	0.5
Zinc and copper	
Zinc and iron	0.2

a) What should be done to the metal rods before connecting them in the cell?

b) State two factors which the students would have kept the same during their experiment.

c) Complete the results table above by predicting the average voltage reading which could have been obtained using zinc and copper rods.

Solution

a) The surface of the rods has to be **cleaned using emery paper and then washed**.

b) You can write almost anything reasonable here except 'use the same metal rods' because the investigation is aimed at finding how different metals affect the voltage! Possible answers include the **temperature of the electrolyte**, the **electrolyte** used, the **concentration of the electrolyte**, the **depth of immersion** of the rods, the **volume of electrolyte** used or the **separation of the rods**.

c) This isn't really a PPA question – it's a bit of problem solving. The answer depends on the positions of the metals in the ECS. The order in which they appear is:

zinc

iron

copper

The zinc/iron cell gives 0.2 V and the iron/copper cell gives 0.5 V. So it's a fair bet that the zinc/copper cell will give the sum of these, **0.7 V**.

Example 10

In the PPA 'Factors affecting voltage', a student investigated the effect of changing the electrolyte.

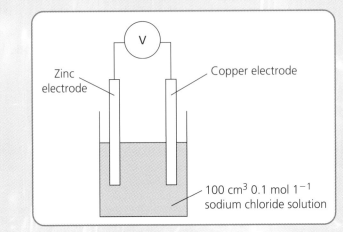

a) Draw and label the cell which would be used to measure the voltage produced when dilute hydrochloric acid is used as an electrolyte.

b) For each cell, two voltage readings were taken. What should have been done before taking the second reading?

Solution

a) At first sight this looks ridiculously simple. But watch out! If you're investigating the effect of changing the electrolyte then this means that all other factors must remain the same. So the electrodes must be identical, and so must the volume and concentration of the electrolyte. The only thing you can change is the electrolyte itself. So what you must draw is:

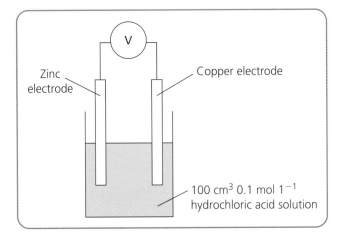

You need to be especially careful that you say '**100 cm³ of 0.1 mol l⁻¹ hydrochloric acid solution**' because if you miss out the volume or the concentration you might end up missing out on the mark.

b) You need to remember that you have to **clean the rods** by sanding them and washing them before every measurement. Just as in the previous example!

Example 11

The PPA 'Reactions of metals with oxygen' was carried out using the apparatus shown below. Three different metals were used.

a) Complete the table to show what would be observed when zinc was used.

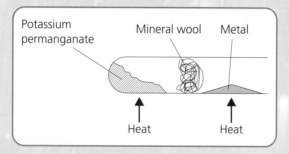

Metal	Observation
Copper	Dull red glow
Magnesium	Very bright white light
Zinc	

b) What was the aim of the experiment?

c) Two safety precautions are wearing safety glasses and making sure the mouth of the test tube is not pointing at anyone. State one other safety precaution which must be taken when using magnesium.

Solution

a) How the metal will behave depends on its position in the ECS, viewed as a reactivity series. Because zinc is above copper and below magnesium in the series, we'd expect its reactivity to be in between as well. So anything in between 'dull red glow' and 'very bright white light' will do. **'Red glow'**, **'bright red glow'** or even **'bright white light'** would do.

b) The aim of the experiment is to **place the three metals in order of reactivity** by comparing the ease with which they react with oxygen.

c) The main precaution is that the experimenter **should not look directly at the magnesium** when it's burning to avoid harm to vision.

LAST WORD

On the Day – Advice and Reminders

Make sure you know the date, time and place of the exam – preferably before the exam! Arrive 15 minutes early and bring with you two pens, two HB pencils and sharpener, a ruler and calculator – one that you know how to use, that is in working order and with fresh batteries. Don't bring your mobile phone. Don't bring your Data Booklet either – one will be provided.

Tackle the multiple-choice part (Section A) first. Don't leave out any questions. Indicate an answer using a horizontal line in HB pencil. Make sure you note, on page 2 of the exam booklet, how to make changes to multiple-choice answers.

Read the questions carefully and make sure you're taking in what the question is asking. Be very careful with questions that contain the words 'both' and 'neither' / 'increase' and 'decrease' / 'greatest' and 'least' to make sure that you really understand what the question is asking you to do.

In Section B, again it's a good idea to read the questions. Start with the first question and work through the rest. If you meet one you can't do then skip it and return to it later – but make sure that you **do** return to it.

Show as much working as you can in calculations. Try to let the marker see what thought processes you've been using. Be careful with the units of your answer – check the wording to see if they are given in the question. If so, it's not necessary to give them. Make sure that your answers are reasonable – for example, if you are asked to work out the number of atoms or molecules in a certain mass of an element or compound then the answer is bound to be a very large number. If you are asked to work out the mass of an atom or molecule then you should expect to get a very small number.

Use a pencil and ruler if you're asked to draw or complete a diagram. Try to make it neat. Use real, recognisable apparatus in your drawing. It's best sticking to cross-sectional type diagrams, rather than trying an artistic three-dimensional drawing. Cross-sectional diagrams are easier to do and are less confusing to interpret. Make sure that you don't close any tubes that are meant to be open.

Don't be put off by questions which seem totally unfamiliar to you (unless you've done no revision, in which case they will all be totally unfamiliar!). Your paper will contain problem-solving questions (often with lots of words) involving reactions and techniques that you haven't seen before. Take time, read the question a couple of times and try to think it through.

Good luck!